WEATHER TRACKER

Backyard Meteorologist's LOGBOOK

Leslie Alan Horvitz

BARRON'S

Leslie Alan Horvitz is an author and journalist specializing in science. His books include *Night Sky Tracker*, *The Complete Idiot's Guide to Evolution*, and *Eureka!: Scientific Breakthroughs that Changed the World*. He lives in New York City.

First edition for the United States and Canada published in 2007 by Barron's Educational Series, Inc.

All inquiries should be addressed to:
Barron's Educational Series, Inc.
250 Wireless Boulevard
Hauppauge, NY 11788
http://www.barronseduc.com

ISBN-13: 978-0-7641-3569-9
ISBN-10: 0-7641-3569-4
Library of Congress Control Number 2005939108

A Marshall Edition
Conceived, edited, and designed by Marshall Editions
The Old Brewery
6 Blundell Street
London N7 9BH
U.K.

Publisher: Richard Green
Commissioning Editor: Claudia Martin
Art Direction: Ivo Marloh
Editor: Johanna Geary
Layout: Alchemedia
Illustrations: Mark Franklin
Indexer: Richard Bird
Production: Nikki Ingram

Printed and bound in China by Midas Printing International Limited
10 9 8 7 6 5 4 3 2 1

Contents

How Weather Works **4**

The Weather Machine 6

Extreme Weather 28

Weather Forecasting 42

Making Weather Predictions **54**

Setting Up Your Own Weather Station 56

Predicting Weather from Clouds 70

Predicting Precipitation 84

Using Atmospheric Pressure and Winds 96

Making Use of Your Resources 104

The Weather Logbook **112**

Glossary 218

Further Resources 221

Index 222

HOW WEATHER WORKS

Weather is what everybody talks about but no one can do anything about. Weather is news. There are cable television stations and Internet sites devoted to nothing but weather. On nearly a daily basis, we check the weather reports to find out whether to bundle up or take an umbrella. But in recent years, with alarm growing over the prospect of unprecedented climatic changes, people have become even more interested in the weather. Weather reports begin to assume the significance of auguries. Does a long, hot summer portend global warming or is it just an aberration that will correct itself next year?

Weather is not the same thing as climate, but weather constitutes a climate. "Weather" describes the meteorological conditions at a particular time and place: the atmospheric pressure, temperature, wind velocity, cloud formation, and precipitation. The term "climate," by contrast, encompasses average and extreme weather conditions over the long term. Different parts of the planet have their own climate. The northeastern United States, for instance, has a temperate climate and the Amazon Basin a tropical climate. Since meteorological conditions at any given time or place influence conditions at other times and at other places, weather should be thought of as an intricate system—or as a machine. This book cannot teach you how to change the weather, but it will help you to understand how weather changes and to pick up some skills that you can use to predict what kind of weather the machine is likely to churn out next.

The Weather Machine

All weather phenomena are produced as a result of both the Sun's heat and atmospheric conditions on Earth. Sunlight is the energy that keeps the weather system in motion, but how that energy is distributed—which regions will receive and retain heat and which regions will obtain only a frugal amount—depends on how sunlight is affected by the atmospheric layers that blanket our planet.

As everybody knows, weather can come in all sorts of forms. Rain, hail, fog, sunshine, winds, temperature—these are all phenomena that constitute weather. All these different kinds of weather have several things in common. First, all weather depends on atmospheric conditions at a given place and time. Second, all weather can change suddenly, which can be attested to by anybody who has ever been caught in a summer thunderstorm. Third, all weather systems are characterized by defined cycles. Fourth, weather requires two vital processes: one involving heat and the other involving water.

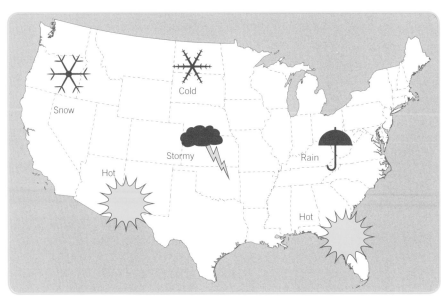

Above: *In this weather map of the United States, weather conditions are shown by simple symbols. The map is a snapshot of weather at a given time but offers no information about the movement of high- and low-pressure systems that will influence weather in the future.*

Temperature

Weather on Earth could not happen without the heat from the Sun. The Sun emits energy at a nearly constant rate but that energy makes itself felt in very different ways depending on the location on Earth in relation to the Sun. Close to the equator, where the Sun is directly overhead, the Earth remains warm all year-round, but much less energy from the Sun ever reaches the poles.

Heat, which is stored in the Earth's atmosphere, is dispersed unevenly over the Earth's surface, resulting in differences in temperature from one place to another, even in locations that are close to one another. Temperature is a measure of the degree of hotness of the air. Temperatures are by and large highest in the tropics and lowest at the poles. It is warmer at lower elevations and colder in mountain ranges. These temperature differences are responsible for differences in density and pressure of air, which in turn produce winds.

Wind

A packet of warm air will expand, making it less dense than the air around it. This will make it more buoyant and it will rise. The rising air will be replaced by air flowing in from surrounding areas, particularly from above cooler surfaces where the air is cooling, getting heavier, and sinking. Wind can be defined as the horizontal movement of air. Winds are named for the direction from which they originate—for example, southerly winds come from the south. On average, winds will blow at speeds of 5 to 15 mph (8–24 km/h), but during hurricanes and typhoons they can attain speeds of 200 mph (320 km/h).

The Water Cycle

Water, like heat, moves between Earth and the atmosphere. This process is known as the water cycle, which gives birth to such phenomena as dew, cloud, and rain. Water on the Earth's surface evaporates into the atmosphere as water vapor. Humidity is a measure of the amount of water vapor in the air. As the temperature rises, air can accommodate more water vapor, doubling its capacity for each temperature increase of 18°F (10°C), which is why humidity is intense on sweltering summer days.

As temperatures cool, the excess water vapor in the air condenses into liquid droplets or ice crystals, forming clouds. As they move, the droplets or crystals bump into one another and join together. When they grow large enough, the droplets or crystals become too heavy to stay suspended in the air and they fall to the ground in the form of precipitation. Depending on the temperature in the air and on the ground, precipitation can descend as rain, freezing rain, snow, hail, ice pellets, and sleet.

Atmospheric Pressure

There is one more vital component of the weather machine: pressure. Atmospheric pressure is defined as the force of the air on a given surface divided by the area of that surface. In places where the air is rising, there is less pressure pushing down on the surface. Where the air is sinking, it produces an area of higher pressure. Low-pressure areas are breeding grounds for storms, so that if pressure is falling rapidly you are likely to be in for bad weather. If pressure is rising rapidly, weather conditions are likely to improve.

The Atmosphere

The Earth's atmosphere consists of five layers extending almost 600 miles (960 km) into space, but nearly all weather "happens" in only one layer, called the troposphere, which begins on the Earth's surface and reaches only to a height of about 7 miles (12 km).

All planets have an atmosphere, or an envelope of gases, surrounding them. The Earth's atmosphere is made up primarily of nitrogen (78 percent) and oxygen (21 percent), water vapor (0–4 percent) and trace amounts of other gases (1 percent), including carbon dioxide, ozone, methane, and nitrous oxide. The atmosphere can be thought of as a blanket that keeps heat from escaping, radiating much of it back to Earth, while protecting us from the harmful effects of the Sun's rays. Without an atmosphere, life would not be possible.

Layers of the Atmosphere

Earth's atmosphere consists of five layers: the troposphere, stratosphere, mesosphere, thermosphere, and exosphere. Confusion can result from different classification systems and nomenclature. For instance, the troposphere and stratosphere are also known as the lower atmosphere, whereas the mesosphere and thermosphere are known as the higher atmosphere. Some classification systems omit the exosphere (from the Greek "exo," meaning out), which is the outermost layer, where the atmosphere thins into space.

The layers are based on temperature changes at varying elevations. However, there are no definitive boundaries separating one layer from the other or the atmosphere from outer space. Together, the troposphere and stratosphere contain 99 percent of all "air" (nitrogen, oxygen, and other gases) in the atmosphere.

The troposphere is often referred to as "The Weather Zone" as it is where most weather takes place. It begins at ground level and extends around 7 miles (12 km). Most of the clouds in the sky are found here. The troposphere, with its high levels of oxygen, is the shell in which life flourishes; in the atmospheric layers above it, conditions are hostile to life. Nonetheless, those layers still play a vital role in shielding us from the Sun's radiation as well as chunks of rock and space junk that the atmosphere incinerates before they can plunge to Earth.

The stratosphere extends some 7 to 30 miles (12 to 50 km) above the Earth's surface. It is drier and less dense than the troposphere and it is also more stratified in temperature, with warmer layers higher up (due to the exposure to ultraviolet energy from the Sun) and cooler layers farther down. (By contrast, the troposphere is cooler higher up and warmer farther down.) The stratosphere's temperature reaches 26°F (-3°C). The stratosphere is where the protective ozone layer is found, which absorbs and scatters harmful ultraviolet radiation.

The mesosphere extends 30 to 50 miles (50 to 80 km) above sea level. With no ozone to absorb radiation, temperatures can fall to -136°F (-93°C). It is the coldest of all atmospheric layers, and any water vapor will freeze into ice clouds.

The thermosphere extends about 50 to 430 miles (80 to 700 km) from the

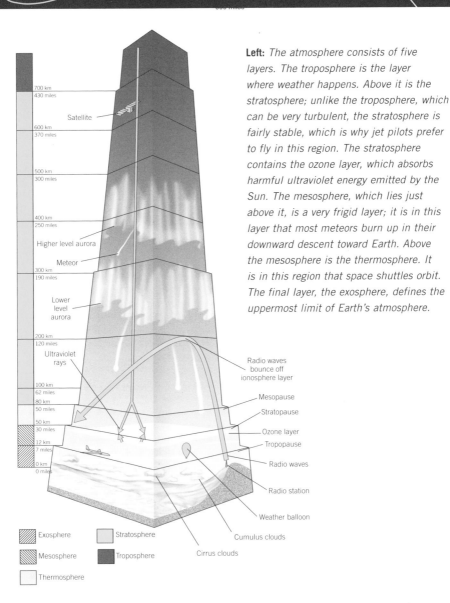

Left: *The atmosphere consists of five layers. The troposphere is the layer where weather happens. Above it is the stratosphere; unlike the troposphere, which can be very turbulent, the stratosphere is fairly stable, which is why jet pilots prefer to fly in this region. The stratosphere contains the ozone layer, which absorbs harmful ultraviolet energy emitted by the Sun. The mesosphere, which lies just above it, is a very frigid layer; it is in this layer that most meteors burn up in their downward descent toward Earth. Above the mesosphere is the thermosphere. It is in this region that space shuttles orbit. The final layer, the exosphere, defines the uppermost limit of Earth's atmosphere.*

Satellite

Higher level aurora

Meteor

Lower level aurora

Ultraviolet rays

700 km
430 miles

600 km
370 miles

500 km
300 miles

400 km
250 miles

300 km
190 miles

200 km
120 miles

100 km
62 miles

80 km
50 miles

50 km
30 miles

12 km
7 miles

0 km
0 miles

Radio waves bounce off ionosphere layer

Mesopause

Stratopause

Ozone layer

Tropopause

Radio waves

Radio station

Weather balloon

Cumulus clouds

Cirrus clouds

Exosphere

Mesosphere

Thermosphere

Stratosphere

Troposphere

surface of Earth. Temperatures can reach above an estimated 2,730°F (1,500°C), because of intense exposure to the Sun's heat. Because of the heat, chemical reactions occur more frequently—and more quickly—here than they do in lower atmospheric layers, causing many atoms to become ionized. (Ionization occurs when an atom acquires an electric charge from the gain or loss of an electron.) That is why this region (together with the mesosphere) is also sometimes called the ionosphere.

The exosphere is the final layer, a transition zone between Earth and space, with no well-defined upper boundary.

The Seasons

Seasonal changes have nothing at all to do with the distance between the Earth and the Sun. The progression of the seasons depends on the tilt and rotation of the Earth in relation to the Sun. As the Earth orbits the Sun, each hemisphere takes a turn at being tilted toward the Sun—this is summer. During the summer, the Sun is higher in the sky and its rays strike the Earth at less of an angle and therefore with greater intensity, bringing warmth.

The Earth's Tilt

You can think of the Earth as a spinning top, but a top that, instead of spinning with its axis vertical, is tilted to one side. In fact the Earth is tilted on its axis by an average of 23.5 degrees (the tilt varies from about 22 degrees to 24.5 degrees). With the Earth tilted to one side, the angle at which the Sun's rays hit any particular spot on Earth changes as the Earth maintains its elliptical orbit around the Sun. You can observe this fact for yourself:

take note that, through the year, the Sun doesn't always rise in the same spot. In the winter in the northern hemisphere, the Sun will rise in the southeast and, as the weeks go on, it will appear to edge slowly to the north, so that by the first day of spring, the vernal equinox, the Sun will rise directly in the east. The Sun appears to continue on its northerly path until the first day of summer, the summer solstice, when it will rise in the northeast. From then, the sunrise can be observed making

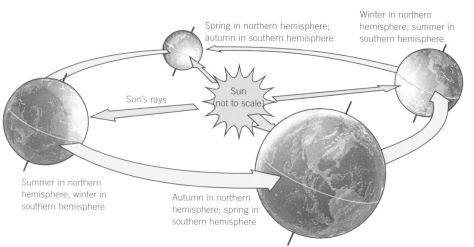

Spring in northern hemisphere; autumn in southern hemisphere

Winter in northern hemisphere; summer in southern hemisphere

Sun's rays

Sun (not to scale)

Summer in northern hemisphere; winter in southern hemisphere

Autumn in northern hemisphere; spring in southern hemisphere

Above: *The Earth is tilted on its axis 23.5 degrees from the vertical. As the northern hemisphere tilts toward the Sun during the Earth's orbit, that hemisphere experiences summer. When the southern hemisphere is tilted toward the Sun, it is the beneficiary of more sunlight.*

The Equinox: Why Day and Night Are Not Equal

The vernal (spring) equinox occurs on or around March 21, when the Sun is in the sky for exactly 12 hours. The same holds true for the autumn equinox, which falls on or around September 21. (The word "equinox," derived from the Latin, means equal nights.) However, the day actually lasts longer than night during the vernal equinox. That's because sunrise takes place when the top of the Sun is seen on the horizon (and not the center). But in fact, at sunrise the Sun is still below the horizon. Its light is refracted (or bent) by the Earth's atmosphere so that it looks as if the Sun has reached the horizon a few minutes before it actually does. For the same reason, sunlight lingers a few minutes longer after sunset. As a result, day is longer than night on the equinox even though in theory day and night should be equal.

a slow advance to the south, rising directly in the east on the first day of autumn, the autumn equinox, and reaching its most southerly point on December 21, the winter solstice. Then the cycle begins again.

Changing Sunlight

The Earth's tilt affects how high the Sun is in our sky. During winter in the northern hemisphere, that hemisphere is tilted away from the Sun, so the Sun does not rise very high in the sky. The opposite is the case in summer. In the southern hemisphere, of course, the seasons are reversed. The Sun's rays strike the Earth more directly the higher it is in the sky, which is why summer is hot and winter is cold. The sunlight's angle affects its intensity in another way, too. The atmosphere filters out or scatters some of the Sun's rays, so, when the Sun is overhead, more sunlight penetrates through the atmosphere because it has less of a distance to travel before reaching the ground.

With the Sun higher in the sky, daylight lasts longer. Close to June 21, the summer solstice, the Earth is tilted such that the Sun is positioned directly over the Tropic of Cancer at 23.5 degrees north latitude, putting the northern hemisphere in a more direct path of the Sun's energy. Meanwhile, in the southern hemisphere, the lower angle of the Sun produces cooler weather and shorter days. In the mid-latitudes, daylight ranges from about 15 hours around the summer solstice to about 9 hours around the winter solstice when the Sun is positioned directly over the Tropic of Capricorn at 23.5 degrees south latitude. That's when it is the southern hemisphere's turn to benefit from more direct sunlight and longer days. At the equator, the days and nights remain roughly the same length all year.

The Water Cycle

The Earth's water is continuously moving through a cycle characterized by three processes: evaporation, condensation, and precipitation. Water rises as vapor from the Earth, condenses into clouds in the form of water droplets and ice pellets, before falling back to the Earth as precipitation.

Understanding the Cycle

Earth and its atmosphere are continuously exchanging water in various forms. This exchange, called the water cycle or the hydrologic cycle, is the only way that Earth is assured of obtaining the fresh water it requires to sustain life.

Heat from the Sun causes water to move from its liquid to its gaseous state: it evaporates from the oceans, rivers, lakes, trees, and plants as water vapor— a phenomenon we know as evaporation. As the water vapor cools, it returns to its liquid state and forms clouds (or fog if it is close to the ground), a process called condensation. Eventually the water returns to Earth by precipitation. Then the rainwater or run-off from melted snow is carried back to the oceans by rivers, streams, and underground tributaries.

The amount of water that is involved in this process is almost unimaginable: some 4 trillion gallons (15 trillion liters) of water fall to Earth as precipitation every day. (And there is yet more water in the atmosphere: above the United States alone, clouds are holding a reserve of some 40 trillion gallons (150 trillion liters) of water every day. Of those 4 trillion gallons (15 trillion liters), about 1.5 trillion gallons (5.7 trillion liters) are absorbed by the ground or flow into rivers, lakes, and oceans. Most precipitation—about 77 percent—falls into the sea. The water that soaks into the Earth nourishes the plants and the trees. The rest of the water is immediately lost as evaporation.

Around 70 percent of Earth is covered by water. There are a total of 326 million cubic miles of water (more than 326,000,000 trillion gallons/ 1234,000,000 trillion liters) on Earth and in its atmosphere. Ironically, with all this water, only a tiny amount of it—1 percent—is present in the rivers, lakes, and groundwater that provide sources of drinking water. Most of the rest—97 percent—is in the oceans.

What form precipitation assumes depends on the temperature of the atmosphere where it originated, the temperature of the air through which it falls, and the temperature of the surface on which it lands. It can be a solid (ice), a liquid (water), or a gas (water vapor). Ice can change on its way to the ground to water or water vapor, or vice versa. If water vapor loses heat, it condenses, becoming liquid again. If liquid loses heat, it freezes and turns into ice.

The water cycle has been going on for 4 billion years, since the formation of oceans on Earth. It took some time for the water cycle to get going for the simple reason that there was no water to be had on Earth when it was formed some 4.5 billion years ago. It is believed that water originated on our planet only when it was sufficiently warmed by the heat of the Sun for hydrogen and oxygen atoms to combine, producing H_2O.

Building a Model of the Water Cycle

You can demonstrate the water cycle by building a simple model requiring nothing more complicated than two half-gallon (2-liter) jars, a rock, masking tape, and food coloring.

1 Pour about 1.5 inches (4 cm) of water into one of the half-gallon jars. Add a few drops of food coloring. This water will represent the ocean.

2 Now place a rock in the middle of the jar. Make sure it is large enough to protrude above the water. The rock represents land.

3 Invert the second jar and place it over the first jar. Tape the two jars together.

4 Place the model on a windowsill where the Sun can reach it. See what happens.

The Sun will heat the colored water in the bottom of the jar. Eventually some of this water will evaporate into water vapor (too small for you to see), which will rise as the air warms. When the water vapor comes close to the sides of the jar, which are cooler, it will lose heat and condense onto the jar. The condensing droplets will not be colored, unlike the water in the bottom of the jar.

Residue of the food coloring is left in the "ocean," as salt is left in the real ocean, so the water will appear more deeply colored. The rock will be more prominent as there will be less water in the bottom of the jar. Eventually the condensed droplets will run back into the ocean.

Below: *The water cycle, upon which practically all life depends, is based on three processes: evaporation, condensation, and precipitation. Water in lakes, streams, and oceans or on the ground evaporates into water vapor (1); as it rises into the atmosphere the water vapor cools and condenses into water droplets or ice, forming clouds (2). Clouds produce precipitation, which returns the water to Earth, where it runs off the land or sinks into the ground.*

Atmospheric Pressure

When a weather forecaster refers to high or low pressure he or she is talking about high or low atmospheric pressure. Atmospheric pressure is the weight of air at a particular location. At sea level, atmospheric pressure is greater than it is at the top of a mountain. Increases and decreases of air pressure play a crucial part in determining weather conditions.

What is Atmospheric Pressure?

Air can be compressed, which is the case when you put air into your tires. When air is compressed it is under higher pressure, as its molecules are more densely packed. Atmospheric pressure is the pressure on any portion of the Earth's atmosphere exerted by the weight of air above it. More technically, it is the force exerted per unit area. It seems odd to think of air as having weight, since under ordinary circumstances you do not feel it bearing down on you as you would if you were carrying a backpack. All the same, air is composed of molecules that, while invisible to the naked eye, do have weight. The pressure at sea level is given in terms of Mean Sea Level Pressure, or MSLP or SLP. At higher elevations, there is less pressure exerted by the atmosphere above, so atmospheric pressure decreases.

At sea level, Earth's atmosphere bears down with a force of 14.7 lbs per square inch (1 kg per square cm). Scientists call the atmospheric pressure at sea level one atmosphere (atm)—the

Popping Ears

Anyone who has ever flown in a plane or climbed a mountain has experienced a change in air pressure. As you climb or descend, you will usually experience a popping in the ears. The pop you hear is the air suddenly passing through the middle ear to balance the external air pressure with the pressure inside your head. At high elevations, you will breathe with more difficulty than at sea level, due to the thinner air. At altitudes of over 8,000 feet (2,500 m) you are likely to experience altitude sickness due to the lack of oxygen. This does not happen on airplanes because the air is artificially pressurized. Airplanes commonly cruise above 18,000 feet (5,600 m), where the supply of oxygen is so small that you would begin to feel ill very quickly. Air pressure outside is less than half of that at sea level. The cabins of airplanes are pressurized to about 75 percent of the air pressure at sea level.

Anticyclones and Cyclones

Anticyclones are characterized by a descending movement of air and a relative increase in barometric pressure over a given area of the Earth's surface, making them harbingers of fair weather. They are the opposite of cyclones such as hurricanes or typhoons, which cause barometric pressure to drop precipitously in a tropical depression (see pp. 34–5). In the northern hemisphere, anticyclones move in a clockwise direction; in the southern hemisphere, they move in a counterclockwise direction. Anticyclones form when colder air from the poles moves toward the equator, set in motion by the Earth's rotation, causing winds called the polar easterlies. Since the air is descending, it is warming and drying.

basic unit for atmospheric pressure. This is equal to the amount of pressure exerted by water at approximately 33.9 feet (10.3 m). That means that a diver at a depth of 33.9 feet (10.3 m) of fresh water is under a pressure of approximately 2 atmospheres (1 atm of air and 1 atm of water).

Measuring Air Pressure

Atmospheric pressure is often measured with an instrument called a mercury barometer, which explains why barometric pressure is used as a synonym for atmospheric pressure. Barometers measure the current air pressure at a particular location in inches of mercury (Hg) or in millibars (mb). In a barometer, mercury settles in its measurement tube at a height dependent on atmospheric pressure. A measurement of 1 atmosphere, the standard pressure at sea level, is equal to 29.92 inches (76 cm) of mercury (1013.25 millibars).

TV weather reports usually give atmospheric pressure in inches of mercury, but meteorologists prefer to work in millibars. Barometric readings on weather reports all refer to the pressure at sea level and not the actual local atmospheric pressure, which may differ with variations in altitude.

Highs and Lows

Air pressure is not just dependent on altitude. It is also dependent on the movement of air masses. Where air is rising, the air pressure is lower; where air is sinking, the pressure is higher. The highest recorded atmospheric pressure, 1086 mb or 32.06 inches (81 cm) of mercury, was reported in Mongolia on December 19, 2001. The lowest recorded atmospheric pressure (not related to tornadoes, see pp. 32–3) was 870 mb or 25.69 inches (65 cm), which occurred during a typhoon in the western Pacific on October 12, 1979.

For meteorologists and other researchers who study weather and climate, variations in pressure are important because they have a direct impact on weather conditions, a subject we will take up in the following pages.

Air Masses

An air mass is a large body of air with similar temperature and humidity that can extend over thousands of square miles. Its uniform nature is due to the fact that it has been formed over one region of the Earth. Air masses formed at the poles, for example, are cold, while air masses formed over an ocean are moist. The movement of air masses is often responsible for bringing abrupt changes in weather conditions.

Air masses are described by their temperature and moisture properties. Typically, air masses are classified according to the characteristics of the region in which they originated. The best source regions for air masses are over large, flat areas of land or water. Those are the areas where air is more likely to stagnate for long enough to take on the characteristics of the surface below. Source regions are classified as having one of three temperature attributes: tropical, polar, or arctic. Air masses have two moisture characteristics: continental or maritime, meaning that they originate over land or sea.

Air masses can be classified in five ways: continental arctic (cA), continental polar (cP), maritime polar (mP), maritime tropical (mT), and continental tropical (cT). Air masses do not remain in their place of origin, but move about, usually in response to winds in the upper atmosphere.

Continental arctic air masses are characterized by extremely cold temperatures and very little moisture. They originate north of the Arctic Circle and primarily impact on Canada and the United States in the winters but rarely form in summer because of the warming of the Arctic by the Sun. All arctic air masses are dry because they have low absolute humidity and undergo very little evaporation in their passage over frigid polar oceans. Continental antarctic air masses are intensely cold at their origin but are modified as they meet warmer air masses over water, so they are known as maritime polar.

Continental polar air masses are characterized by cold and dry air, but are not quite as cold as the air masses formed over the Arctic because they originate farther to the south. These air masses are often responsible for the kind of weather experienced by the United States during the winter. They can also form in summer, when they bring clear weather to the northern United States. In Europe these air masses originate over the Eurasian land mass north of 50°N and east of 25° and are exclusively a winter phenomenon; they bring very cold conditions characterized by clear skies and severe frost.

Maritime polar air masses are cool and moist and often bring cloudy, damp weather with them. They originate over the northern Atlantic and Pacific oceans as well as oceans of antarctic regions. They can form at any time of year and are more temperate than the continental polar air masses. Northwesterly winds produced in the arctic regions of Greenland and northern Canada carry these air masses to western Europe. In Asia these air masses play only a minor role in the air cycle.

Above: *When maritime tropical air masses from the Caribbean and Gulf of Mexico move into the southern United States, they can be accompanied by intense precipitation. This region is prone to tropical storms from June to November.*

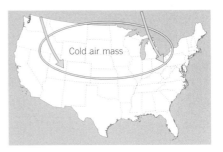

Above: *In winter, continental polar air masses creep down into the northern tier of the United States from Canada. If conditions are right, these air masses can sit for several days, bringing colder temperatures, blustery winds, and snow.*

Maritime tropical air masses are characterized by warm temperatures with plentiful moisture. They originate over the warm waters of the southern Atlantic Ocean and the Gulf of Mexico and influence the weather in the eastern United States. Although these air masses can develop at any time, they are most prevalent in the United States during the summer, when they are responsible for hot, humid days across the south and east. The air is usually carried over to the British Isles on southwesterly winds. As it passes over the cooler waters of the Atlantic, the air becomes close to saturation. In the west of the British Isles, the weather generated by these air masses is characterized by low cloud cover, drizzle, and hill fog. In Asia, maritime tropical air is usually found along the eastern coast of China and over Japan during the summer, where its effect is similar to that observed in North America.

Continental tropical air masses are hot and dry. In the United States they usually form over the deserts of the southwest and northern Mexico during summer and can produce blistering heat in the midwest. These air masses almost never occur in winter. In Europe, these air masses are observed during the summer when easterly and southerly winds draw hot, dry air from North Africa. In Asia these air masses play only a minor role in the air cycle.

Movement of Air Masses

Over the course of their journeys, air masses are influenced by the different air masses and surface conditions they encounter as they move. So a continental polar air mass born in Canada will begin to warm if it passes over land with warmer temperatures. When air masses collide, usually in middle latitudes, they can create some interesting weather.

This interesting weather occurs at the boundary, or interface, between the two air masses. At that boundary, the temperature differences between the clashing air masses becomes intensified. Meteorologists call that area of intensification a frontal zone or simply a front. For a full explanation of fronts, see pp. 22–5.

Wind and Atmospheric Circulation

Wind is air in motion produced by an imbalance between atmospheric regions of different pressure. What direction the wind will take, how fast it will blow, and for how long, depend on a number of variables, including the rotation of the Earth. The winds keep weather circulating and changing.

What is Wind?

Wind is the movement, sometimes fast, sometimes slow, of air molecules from one place to another. Air is in constant motion because it is always trying to find equilibrium between areas of more air molecules (higher pressure) and those with fewer (lower pressure). Think of a vacuum-packed container—that is a container from which air has been removed, meaning that it is a low-pressure environment. When you open the container, the satisfying whoosh you hear is the sound of air from the outside (a high-pressure environment) rushing into it. That same phenomenon, on a much greater scale, is played out on a global level. It is known as the pressure gradient (also known as the barometric gradient), which refers to the change in atmospheric pressure over a given distance at a given time.

The movement of wind—what direction it takes and the speed at which it blows—depends on three forces: the pressure gradient, the Coriolis force, and surface friction. The Coriolis force is the deflective effect of the Earth's rotation on all free-moving objects, including the atmosphere and oceans. It causes wind to move toward the right of its motion in the northern hemisphere and toward the left in the southern hemisphere. Surface friction is the resistance to the movement of air as it flows across the ground.

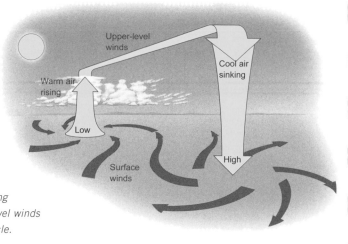

Right: *When air over warmer areas rises, air from surrounding areas rushes in to replace it. This creates surface winds. When the rising air reaches the upper atmosphere, it rushes to replace sinking air, producing upper-level winds and completing the cycle.*

Upper-level winds

Cool air sinking

Warm air rising

Low

Surface winds

High

What is Atmospheric Circulation?

Atmospheric circulation is the large-scale horizontal movement of air, achieved by wind. It plays a vital part in circulating heat from the Sun across the surface of the Earth. The tropics receive far more heat from the Sun than the rest of the globe. If the atmosphere did not exist, the tropics would get hotter and hotter while other regions would get colder. The equator is the driving force behind the global wind pattern. Here, intense heating by the Sun causes the air to heat and rise quickly, drawing in strong, steady winds called the trade winds from each hemisphere. At the Earth's poles, with their vast areas of snow and ice, cold air sinks and spreads outward. When it reaches the Earth's middle latitudes it meets warmer winds along a boundary called the polar front. As the two winds collide they produce the storms called frontal depressions (see pp. 96–7).

Atmospheric circulation is not the only system involved in circulating the Sun's heat: in the same way, the warm tropical waters of the ocean move toward the poles, while the colder waters at the poles move toward the equator.

Below: *Atmospheric circulation creates a pattern of movement in the atmosphere. The flow of air from the equator descends at between 30° and 35° latitude both north and south, in a region known as the horse latitudes, which is characterized by light winds and high pressure. The westerlies are the prevailing winds in the middle latitudes between 30° and 60° latitude, blowing from the high-pressure area in the horse latitudes toward the poles. These winds are predominantly from the southwest in the northern hemisphere and from the northwest in the southern hemisphere. The polar easterlies are the prevailing winds that blow from the high-pressure areas at the north and south poles toward the low-pressure areas of the polar fronts at around 60° latitude. Cold air sinks at the poles, creating high pressure, forcing an outflow of air toward the equator. That outflow is deflected eastward by the Coriolis force.*

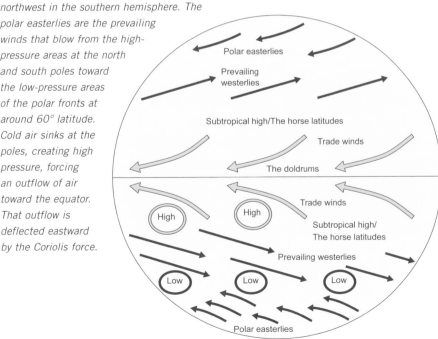

Polar easterlies

Prevailing westerlies

Subtropical high/The horse latitudes

Trade winds

The doldrums

Trade winds

High High

Subtropical high/
The horse latitudes

Prevailing westerlies

Low Low Low

Polar easterlies

Jet Streams

Jet streams are rapidly flowing currents of air high above the Earth's surface. There are two major jet streams—at the polar latitudes in the northern and southern hemispheres—and two minor, subtropical jet streams closer to the equator.

What Are Jet Streams?

The Earth's four jet streams mark the boundaries of regional climates. The regions around 30°N/S and 50–60°N/S are areas where temperature changes are the greatest, and so the strength of the wind increases. The 50–60°N/S region is where the polar jets are located, with the subtropical jets located around 30°N/S.

Jet streams act as the "engines" of weather. They are also called planetary winds because they are part of atmospheric air circulation as opposed to local air circulation on the Earth's surface. These winds, which flow above 20,000 feet (6,000 m), just below the boundary of the troposphere, propel storms and influence high- and low-pressure systems on the Earth's surface.

These meandering currents are typically thousands of miles long, a few hundred miles wide, and only a few miles thick.

Because of the Earth's rotation, jet streams generally flow from west to east around the Earth, with a wind speed of about 92 mph (147 km/h), although in winter, when temperature differences are greatest, they can reach speeds of 300 mph (480 km/h). During winter, flights from the United States to Europe gain added speed because they are pushed along by the northern polar jet stream.

But the speed of a jet stream is not constant: it can accelerate or decelerate or peter out altogether. The strength of the wind depends on variations in temperatures on the Earth's surface, and the sharper the differences, the stronger

Discovery of Jet Streams

The jet streams owe their discovery to the Japanese meteorologist Wasaburo Ooishi, who made observations in the 1920s by tracking weather balloons. Several years would have to pass before his work became known outside Japan, in part because he published his findings in Esperanto, an international language which, however, was not widely understood. In his native country, though, the military used knowledge of the jet streams to launch balloons armed with incendiaries at the United States during World War II, a tactic that had only limited success. The theory of jet streams was eventually developed by Erik Palmén (1898–1985) and other members of what became known as the Chicago school of dynamical meteorologists.

The Sun and the Jet Stream

The Sun's solar activity, which varies over an 11-year cycle, affects weather on Earth. According to recent findings from NASA, an increase in solar activity will create cloudier conditions over the United States. Researchers believe that this is because the northern polar jet stream moves northward and causes changes to climate patterns. Several studies have shown a possible link between a peak in solar activity—known as the solar maximum—and the northward flow of the jet stream.

the jet stream winds. Jet streams arise along the upper boundaries of large masses of warm and cold air. These boundaries are also the locations where weather fronts generally form. If a front is coming through, bringing wind and rain, it is likely that the jet stream may have had a role in creating it.

Those temperature variations also account for changes in the direction of the jet streams. When it is bitter cold in the United States, for instance, the northern hemisphere polar jet stream flows south, following the warm-cold boundary, sometimes moving over the Gulf of Mexico.

In a mild winter or during the summer in the United States, that jet stream may retreat into Canada. Jet streams can affect weather in other ways: they are instrumental in the formation of tropical depressions—a group of thunderstorms that becomes organized into a central system of rotating winds. Meteorologists believe that these systems form around the jet streams and percolate downward.

Jet streams should be distinguished from jet streaks, which occur at lower altitudes, often only a few hundred feet above the ground. Jet streaks are winds that are faster than the surrounding winds.

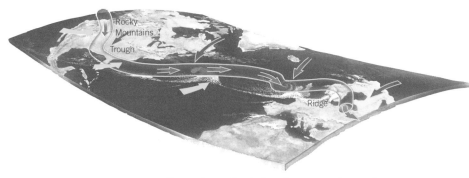

Above: *The height of the Rocky Mountains in North America causes the northern hemisphere polar jet stream to swing southward, often producing a trough (or area of lower pressure) to the east. A ridge is an area of higher pressure.*

Fronts I

When air masses of different temperatures cross paths they produce fronts. These fronts are transitional areas or boundaries and can range from 20 miles (32 km) to 100 miles (160 km) across. Fronts can bring changes in the weather, so forecasters are always on the lookout for them.

The type of front will depend on the direction in which the air mass is moving and on its properties. There are four types of fronts: warm fronts occur when a warm air mass is moving into a colder air mass; cold fronts occur when a cold air mass is moving into a warmer one; stationary fronts occur when the air masses are not moving against each other; and occluded fronts occur where a faster-moving cold front overtakes a slower-moving warm front.

Cold front

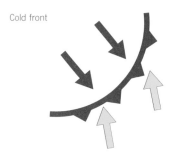

Warm front

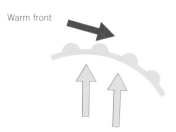

Stationary front

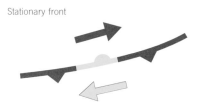

Occluded front

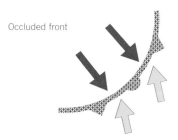

Above: *The four front types can be spotted by these symbols on a weather map.*

Cold Fronts

A cold front is the leading edge of an air mass of colder and drier air than the air whose territory it is about to barge into. It typically moves at 9 to 30 mph (15–50 km/h), faster than a warm front, and tends to produce the most violent weather. As a cold front passes the warmer air mass, it can cause a drop in temperature of about 15°F (9.4°C) within the first hour. As a

cold front moves across a region, it creates turbulence—a vertical movement of air.

Cold fronts herald their advance with the development of cirrus clouds—thin, wispy clouds that form at high altitude (see pp. 72–3)—and as they barrel ahead bring with them a large variety of clouds. They are also characterized by squall lines: a string of strong thunderstorms parallel to and ahead of the front. Precipitation is usually just behind the front. This is the result of frontal lifting, in which the cold air sinks below the warmer air mass, which cools in rising, triggering condensation. A cold front usually brings cooler weather, clearing skies, and a sharp change in wind direction.

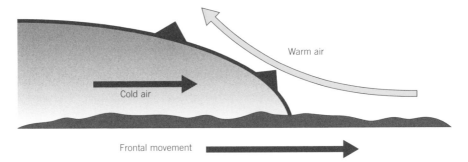

Above: *Warm air rises when it meets a cold front, in a process known as frontal lifting. Cold fronts bring colder, drier air on the leading edge of an approaching weather system. They are associated with storms.*

Warm Fronts

A warm front brings with it air that is warmer and moister than the air ahead of it. Warm fronts tend to occur northeast of a cold front. In a warm front, warm air will rise up and over the cooler, denser air at the surface, a process that leads to cloud formation, saturation, and ultimately precipitation. Predictably, the advance of a warm front will make the air ahead of it warmer and more humid. Warm fronts are much more sluggish than cold fronts. Although they are capable of causing thunderstorms (see pp. 88–9), warm fronts are typically less violent than cold fronts and more frequently produce light to moderate continuous rain to the north of the front as the warm air, driven by low-level southerly winds, rises over cooler, denser air. This lifting results in cloud formation and precipitation. Cirrus clouds about 700 miles (1,000 km) ahead of the front are followed by mid-level altostratus or altocumulus clouds and finally by lower-level stratus clouds and sometimes by fog (see pp. 70–83).

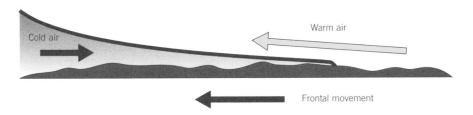

Above: *Warm fronts bring warm, moist air and tend to produce a great many clouds. Although associated with moderate weather, they can also create conditions for violent thunderstorms in spring and summer.*

Fronts II

In addition to cold and warm fronts, meteorologists have identified two other front types: stationary fronts and occluded fronts. The former are often marked by cloudy, rainy weather. Occluded fronts typically occur as a result of a storm system: cold occluded fronts create similar conditions to cold fronts, while warm occluded fronts have a similar effect to warm fronts.

Stationary Fronts

Cold fronts can overtake and overwhelm warm fronts, and sometimes warm fronts can push off cold fronts. However, there are times when fronts come to a halt, resulting in a standoff between competing air masses. Stationary fronts generally form when two air masses collide without either being strong enough to replace the other. Stationary fronts often form as a result of polar air masses that stall; as their temperatures moderate and winds diminish because of the exposure to warmer air, they begin to lose the characteristics that made them polar. Stationary fronts are not entirely welcome since they can frequently cause cloudy, wet weather, with rain or snow, lasting up to a week or more.

A stationary front is a boundary zone keeping the two air masses from colliding. If you were to cross from one side of the boundary zone to the other you would register a perceptible change in temperature and/or a shift in wind direction. For example, temperatures south of the front might be about 50°F (10°C) with winds from the southeast, whereas north of the front, temperatures might be 40°F (5°C) with winds shifting to the northeast. Winds on either side of the front often blow parallel to one another. Warmer air will begin to supplant cool air along the front, bringing with it clouds and precipitation. In that sense, stationary fronts behave more like warm fronts than cold fronts, with the principal difference that they are immobile whereas warm fronts are in constant motion. Once the boundary area begins to move, the stationary front will dissipate and turn into either a warm or cold front.

Occluded Fronts

Occluded fronts form as a result of storm systems. Occluded fronts require three ingredients: a cold front, a warm front, and a cooler air mass. The storm brings both the cold and warm fronts. The cool air must have already been there before the storm arrived. As a cyclone (a thunderstorm, for instance) develops, it is preceded by a warm front, the leading edge of a warm, moist air mass. The faster-moving cold front—the leading edge of a colder, drier air mass—is rotating around the storm. In other words, the warm air is moving ahead of the storm and the cold air is wrapped around the storm. Meanwhile, the cooler air is located to the north of the warm front. As the storm intensifies, the cold front catches the warm front. The cold front does not overwhelm the warm front, however: the two fronts lock

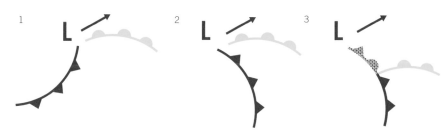

Above: *An occluded front results when a cold front (represented by the solid lines with triangles) overtakes a warm front (represented by solid lines with hemispheres). Typically, a developing storm—a cyclone in meteorological terms—is preceded by a warm front. Meanwhile, colder, drier air is rotating around the storm. The faster-moving cold air eventually catches up with the warm air mass (3), forming a boundary region between the two air masses. This region is an occluded front. The change is indicated in 3 by the blue line, which now has three triangles and one hemisphere.*

heads in effect. The result is an occluded front. (Occlusion means blocked: the fronts cannot move in any direction because they are blocked in.) That front represents the boundary that separates the new cold air mass (to the west) from the cool air mass already present north of the warm front.

There are two types of occluded fronts: a cold occlusion and a warm occlusion. A cold occlusion occurs when the air behind the front is colder than the air ahead of it. The coldest air moves in under the cool air. In other words, the coldest air supplants the less cold air. A cold occluded front acts similarly to a normal cold front. A warm occlusion occurs when the air behind the front is warmer than the air ahead of the front. The cool air (that is to say, the warmer mass of air) will rise over the colder air at the surface. So it follows that a warm occluded front will behave in much the same way that a normal warm front does. In both types of occluded front, however, well-defined boundaries separate the coldest air, the warmest air, and the cool air.

Gust Fronts

Thunderstorms (see pp. 88–9) are frequently preceded by a cool blast of wind just before the rain starts falling. This blast of cool air is caused by the storm's "gust front"—the leading edge of air cooled by rain produced by the thunderstorm. Gust fronts are mini cold fronts capable of causing a drop in temperature of up to 10°F (5°C) in only minutes.

These gust fronts can be very destructive, with winds of up to 100 mph (160 km/h). They are also triggers, because they can spawn new storms at a considerable distance.

Climate

Weather is what happens day to day—rain, snow, fog, fair skies—while climate is weather over the long haul based on an average, and extremes, of such factors as temperature, precipitation, and wind velocity. Climate is a result of natural weather patterns but it can also be influenced by humans. Big cities, for example, are heat-generating machines that can affect the weather patterns and climate.

Weather and climate are interconnected. Since climate refers to average weather over the long term, it is much more predictable than weather. "Climate is what you expect and weather is what you get," meteorologists say. Changes in weather from day to day do not affect climate unless those changes presage a significant change in the pattern of weather (see pp. 28–9).

Climate is determined by two components: the natural and the anthropogenic (caused by humans). In the former case, climate is dependent on the biosphere (which includes the atmosphere and the Earth's surface), the geosphere (the Earth's crust, mantle, and core), and the hydrosphere (all water on and around the Earth). Humans, of course, influence climate by the use they make of resources,

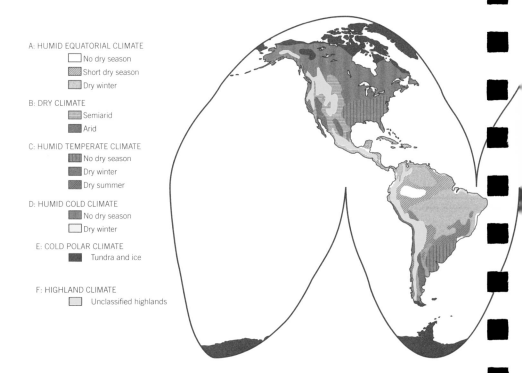

A: HUMID EQUATORIAL CLIMATE
- No dry season
- Short dry season
- Dry winter

B: DRY CLIMATE
- Semiarid
- Arid

C: HUMID TEMPERATE CLIMATE
- No dry season
- Dry winter
- Dry summer

D: HUMID COLD CLIMATE
- No dry season
- Dry winter

E: COLD POLAR CLIMATE
- Tundra and ice

F: HIGHLAND CLIMATE
- Unclassified highlands

which could include agriculture on the one hand and industrialization on the other. Any significant change in the biosphere, geosphere, or hydrosphere can cause changes in climate.

Climate Zones

There is not one climate on Earth but several different climates or climate zones: tropical, desert, temperate, and polar among them. The key cause of differences in climate is the distribution of the Sun's energy, with more heat in the tropics and less at the poles. The seasons (see pp. 10–11) also affect climate. The temperate zones (so called because they are not prone to extremes in temperature or precipitation) in the northern tier of the United States and Europe will experience a seasonal progression, with cold winters and hot summers, while the tropics generally experience wet and dry seasons. Temperatures tend to be higher in the tropics and lowest in the deserts at night and at the poles.

The variation of temperature on the planet is enormous: from over 122°F (50°C) in the desert to below -112°F (-80°C) at the poles. The average global temperature is about 59°F (15°C). Precipitation also differs considerably from one part of the world to another, with higher levels in the tropics and lower in the deserts and at the poles.

Below: *This map displays the range of climates on the planet. A region's climate depends on a variety of factors, including its topography, precipitation, and temperature. Climate tends to remain stable over long periods of time but it can undergo dramatic changes as a result of environmental influences or human intervention.*

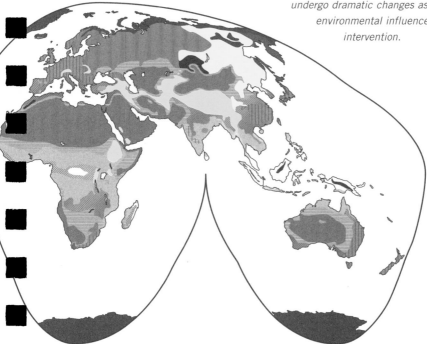

Extreme Weather

Climate change is big news these days as scientists gather evidence showing that the Earth is becoming warmer largely as a result of the build-up of emissions from fossil fuels—a phenomenon known as the Greenhouse Effect. The Greenhouse Effect may also be responsible for an increasing frequency in violent weather, such as hurricanes, tornadoes, floods, and drought.

Of all the problems that confront us, global climate change might be the most important. Although debates still rage among politicians about whether global warming is actually taking place, there is little doubt of its reality among the scientific community. The question for scientists is not whether the climate is undergoing significant change, but how far advanced it is and whether the process can be slowed. Few scientists believe that global warming can be stopped entirely.

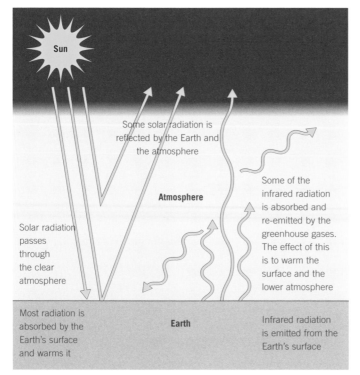

Sun

Some solar radiation is reflected by the Earth and the atmosphere

Atmosphere

Some of the infrared radiation is absorbed and re-emitted by the greenhouse gases. The effect of this is to warm the surface and the lower atmosphere

Solar radiation passes through the clear atmosphere

Most radiation is absorbed by the Earth's surface and warms it

Earth

Infrared radiation is emitted from the Earth's surface

Left: *Because of a build-up of gases in the atmosphere (mainly carbon dioxide), some of the energy that would otherwise escape into space is instead reflected back to the surface of Earth.*

The Greenhouse Effect

Rising temperatures are due to what is known as the greenhouse gases. The Greenhouse Effect is a natural process: gases in the atmosphere prevent too much heat from escaping and making the planet uninhabitable. The cause for concern, though, is that additional greenhouse gases may be heating up the Earth excessively. The amount of certain gases in the atmosphere has increased in recent decades, primarily carbon dioxide (CO^2), but also nitrous oxide, methane, and water vapor. These act like a blanket, trapping the energy. The major culprit is the burning of fossil fuels like oil and coal. The 1998 Kyoto Treaty was intended to curb these gases, but the United States has withdrawn from it, mainly over concerns that its provisions would harm the economy. As China, India, and other developing countries continue to modernize and consume more energy, the prospect for cutting back on CO2 emissions has dimmed even more.

Global Warming

The evidence for global warming is mounting with every passing year. In the lower troposphere, temperatures have increased between 0.2°F and 0.4°F (0.12–0.22°C) per decade since 1979. Some 53 cubic miles (221 cubic km) melted away from the Greenland ice sheet in 2004, compared to 23 cubic miles (96 cubic km) the previous year. The melting of the ice sheets causes the sea level to rise. Since the start of the 19th century, the sea level has been rising at 0.04 to 0.12 inches (1 to 3 mm) per year. As sea levels rise, islands and low-lying bodies of land are increasingly in danger of being flooded. Many South Pacific islands are likely to be submerged entirely, but eventually even a good part of

Manhattan might be under water. And the trend is accelerating: this is because polar ice reflects the Sun, diminishing its heating effect, but the sea absorbs sunlight, so that the water becomes warmer, hastening the melting. Average temperatures in the Arctic have risen 9°F (5°C) since 1968.

Nineteen of the 20 hottest years on record have occurred over the last 25 years. The term global warming is in some ways a misnomer because it does not mean that every part of the planet will grow uniformly warmer. For instance, global warming may change the course of the Gulf Stream (and its extension, the North Atlantic Drift), which brings warm water to the northeastern United States and western and northern Europe from the Caribbean. If the Gulf Stream is to continue to flow northward, it must be balanced out by a comparable, deep return current of cold, dense water from the Nordic seas. Melting of the Arctic ice cap because of global warming is likely to disturb the flow of the return current. That in turn would have the effect of slowing and weakening the flow of the Gulf Stream, preventing warmer waters from reaching northern Europe and producing much colder conditions.

The Ozone Hole

Global warming may also indirectly accelerate the depletion of the ozone layer—the atmospheric blanket that protects Earth against harmful ultraviolet light—by cooling the stratosphere. The ozone hole, as it is known, is caused primarily by chlorofluorocarbon compounds such as Freon, once commonly used in refrigerators. The depletion of ozone will permit more ultraviolet light from the Sun to penetrate the atmosphere and cause an increase in eye disease and skin cancer.

Climate Change

In 2005, New Orleans was battered by Hurricane Katrina, while that year Europe suffered one of its bitterest winters on record. Are these events aberrations or are we seeing the beginning of a new era in which extreme weather is more the norm than the exception? The evidence suggests that global warming is in fact transforming the climate in profound ways that make extreme weather more and more likely.

Ordinarily, western regions of the United States benefit from water produced by melting snow on mountain slopes. But now that spring-like temperatures are occurring weeks earlier, the snow is melting sooner, leaving less for inhabitants to consume later in the summer. Higher temperatures have dried up the soil, too, producing drought-like conditions. On the other hand, North America has experienced some of the wettest winters ever seen. In 1996 the northeastern United States reported its heaviest rainfall in 102 years. That same year, Northern Europe experienced unusually dry conditions. Belgium reported the lowest levels of rainfall since records started in 1833. The U.S. National Oceanic and Atmospheric Administration (NOAA) has reported that the number of blizzards and heavy rainstorms in the United States has jumped 20 percent since 1990. Heavier storm activity has also been occurring in Australia and South Africa.

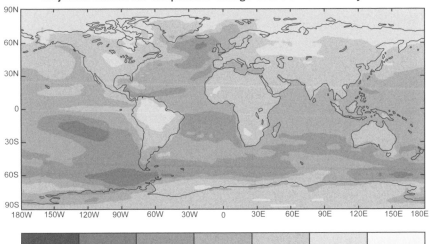

Projected Surface Air Temperature Change Over the Next Century

0°C/0°F   1°C/1.8°F   2°C/3.6°F   3°C/5.4°F   6°C/10.8°F   10°C/18°F

Above: *Constructed by the Climatic Research Unit, University of East Anglia, this model predicts that, between 1990 and 2100, global temperatures may increase by 2.5°–10.5°F (1.4–5.8°C), with the greatest change occurring at higher latitudes. This assumes that emissions of greenhouse gases will double over the 21st century if mid-range economic growth continues.*

Since 1970, ocean waters have warmed by 1°F (0.5°C). While this may not sound like a lot, this rise dramatically increases the incidence of hurricanes and typhoons. In the past 35 years the number of Category 4 and 5 hurricanes (the most devastating) worldwide has doubled, while the wind speed and duration of all hurricanes has jumped 50 percent.

Extreme weather may also be intensifying as a result of a phenomenon known as El Niño periods. A natural phenomenon, El Niño (see pp. 38–9) occurs every three to seven years and is characterized by higher sea temperatures in the western Pacific, causing major changes in global weather patterns. Although scientists are uncertain what triggers these events, there is little doubt that they are occurring with increasing frequency since 1976. And they are doing more damage: the El Niño of 1982–3 was the worst in the preceding 50 years, causing widespread droughts, floods, and hurricanes that killed 2,000 people.

Counting the Cost

Weather extremes exert a terrible toll. In the United States, just between August 1992 and May 1997, 21 weather-related disasters caused 911 deaths and $90 billion worth of damage. In 1998 floods along the Yangtse River in China took 4,000 lives, with economic losses of $30 billion. A heat wave in May of 2002 claimed over 600 lives in India. The effects of extreme weather are often exacerbated by human degradation of the environment. For instance, in 2004, mudslides caused by Tropical Storm Jeanne killed 700 people in Haiti. The trees that could have held the ground in place had been cut down by impoverished people desperate for cooking fuel.

Record Weather Extremes

Highest recorded temperature: 136.4°F (58°C) at Al'Aziziyah, Libya, on September 13, 1922.

Lowest recorded temperature: -129°F (-89°C), at Vostok, Antarctica, on July 21, 1983.

Greatest temperature change in one day: Drop from 44°F (7°C) to -56°F (-48°C), in Browning, Montana, on January 23, 1916.

Most rainfall in a day: 73.6 inches (1.86 m) on La Reunion, Indian Ocean, on March 15, 1952.

Greatest single snowfall: 189 inches (4.8 m) at Mount Shasta Ski Bowl, California, February 13–19, 1959.

Hottest inhabited location: 86°F (30°C) average annual temperature at Djibouti, Djibouti.

Driest inhabited location: 0.003 inches (0.007 cm) annual rainfall, Atacama Desert, Chile.

Fastest wind gust: 231 mph (372 km/h) at Mount Washington, New Hampshire, on April 12, 1934.

Fastest wind in a tornado: 318 mph (511 km/h) at Oklahoma City, Oklahoma, on May 3, 1999.

Heaviest hailstone: 7.5 lbs (3.4 kg) at Hyderabad, India, on March 10, 1939.

Tornadoes

Tornadoes, or twisters, are distinguished by their funnel-shaped column of violently rotating wind. They are among nature's most destructive phenomena. Although most twisters are weak, some have left a trail of death and devastation in regions of the southern and midwestern United States, where they occur frequently in spring and summer.

One of the fiercest types of storms unleashed by nature, a tornado is a rotating column of air that forms in a thunderstorm and extends to the ground. The most powerful tornadoes are capable of wind speeds of over 250 mph (400 km/h). Tornadoes are defined as weak, strong, or violent, but most—7 out of 10—are weak, typified by a thin, ropelike appearance and rotating wind speeds no greater than about 110 mph (177 km). There are about 1,000 tornadoes a year, occurring mainly in the United States. Nonetheless, tornadoes are still rare: about 1,000 thunderstorms occur for every tornado created. Tornadoes can

visit tremendous destruction along their paths, which can exceed 1 mile (1.6 km) in width and 50 miles (80 km) in length, but only the violent forms of these storms are capable of leveling buildings. The Fujita scale rates tornadoes 1 to 6 according to the damage that they cause.

Tornado Formation

Conditions have to favor the formation of severe tornado-spawning thunderstorms. For one thing, the atmosphere has to be unstable, with a warm, moist air mass near the ground and relatively cold and dry air in the upper atmosphere. This will cause the warm air to rise, then cool, condense, and form precipitation-producing clouds. For another, the location should be one where dry polar air masses meet warm, moist tropical air. That is why the south and southwest of the United States are so prone to these storms, earning the sobriquet Tornado Alley. In that region these air masses are likely to meet in spring, which is when most thunderstorms in this region form. Farther north—in the midwestern United States—they tend to form in the fall. That is because the northern plains in the United States take longer to warm.

Thunderstorms can accumulate a vast amount of energy. The energy is created when clouds form from the condensation of warm, moist, rising air. That process

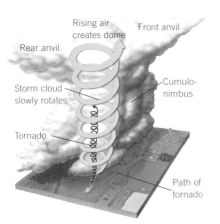

Rising air creates dome

Front anvil

Rear anvil

Storm cloud slowly rotates

Cumulo-nimbus

Tornado

Path of tornado

Above: *Tornadoes can form over sea or land but almost always emerge out of thunderstorms and assume a funnel-shaped violently rotating column of air.*

releases latent heat (the water vapor is giving back to the atmosphere the heat that was taken when the liquid water evaporated into gaseous water vapor in the first place) and as this energy builds up it creates huge updrafts into the cloud.

But not all severe thunderstorms produce tornadoes. A particular kind of thunderstorm called a supercell is needed. A supercell is actually composed of several smaller thunderstorms; one cell may only endure for 20 minutes but a group of cells can last for much longer. A supercell's longevity accounts for why it is capable of causing tornadoes. It gives more time for wind to come into the storm.

Sharp changes of wind speed and/or direction in the lower part of the atmosphere are thought to be responsible for the characteristic rotation of the tornado. As the convective updraft of a thunderstorm draws in air, it changes its direction from horizontal to vertical. That causes the updraft to rotate. As it continues to rotate, it causes clouds to descend inside its center. This spinning central cloud is the funnel cloud that distinguishes tornadoes. The rotating updraft, the spinning column of rapidly rising air—also called a vortex—will create reduced air pressure on the ground below the tornado. It is, in effect, a mini low-pressure system. That very low-pressure area below the tornado sucks more air—and sometimes objects—into it. A tornado's updraft must be accompanied by a downdraft of air cooled by the precipitation in the upper part of the storm. As the downdraft intensifies, it cuts off the updraft, stopping the inflow of moist air, essentially depriving the storm of the fuel it requires to continue. That spells the death of the tornado.

Average Number of Tornadoes per 10,000 Sq Miles (25,000 sq km) Per Year in the United States

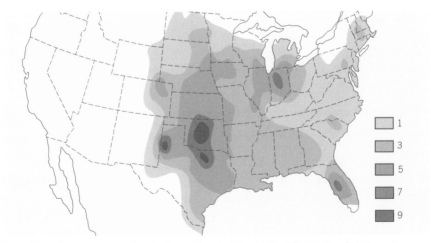

Above: *Tornadoes occur so often in the midwest of the United States that it is known as Tornado Alley. Typically there are 1,000 tornadoes in the United States each year. Tornadoes occur in other parts of the world, including the United Kingdom, which has 15–30 small tornadoes each year.*

Tropical Cyclones: Hurricanes and Typhoons

The powerful storms that develop in summer and autumn mainly in the tropical waters of the Caribbean, south Atlantic, and eastern Pacific oceans, all begin in much the same way: as a system of clouds, strong winds, and high moisture that form around an area of low pressure.

Storm Formation

A "tropical cyclone" is the generic classification for low-pressure systems that develop in the tropics: tropical depressions, tropical storms, hurricanes, and typhoons. These storms are given different names in different parts of the world: hurricanes in the south Atlantic, the Caribbean, the Gulf of Mexico, and off the northwestern coast of the United States; typhoons in the northwestern Pacific Ocean; and cyclones in Australasia, the southwestern Pacific, and the southeastern Indian Ocean. They are all characterized by cyclonic winds—winds swirling around a central eye—and they are all low-pressure systems. (The lowest barometric pressure reading was taken inside a hurricane.)

For hurricane formation, the waters must be sufficiently warm, at least 80°F (27°C). Hurricane formation also requires moist air and converging equatorial winds. A hurricane begins as a thunderstorm, with swirling clouds, wind, and precipitation. If wind speeds reach 38 mph (61 km/h), it will become a tropical depression. When wind speed increases, it becomes a tropical storm (at which point it acquires a name), and if winds intensify further, reaching 74 mph (119 km/h), it becomes a hurricane.

During the process, warm, humid ocean air begins to rise rapidly; the water vapor in the air condenses, forms clouds, and produces precipitation. The condensation releases latent heat, warming the cooler air aloft and causing it to rise. As this air rises, it is replaced by even warmer, more humid air from the ocean. So heat is being removed from the surface and transported into the atmosphere. The exchange of heat creates a pattern of wind that circulates around a center in much the same way that water goes down a drain. The wind pattern is rather chaotic with winds converging from all directions. These winds give an added

Below: *During hurricane formation, hot air over tropical oceans begins to rise, producing tall, mainly cumulonimbus clouds. With the depletion of air at the ocean surface, a low-pressure system forms, which sucks in air from the surrounding area. That air is warmed in turn, rises, and produces more tall clouds. The clouds begin to join in spiral bands, generating thunderstorms. Higher atmospheric winds act to keep the storm organized.*

How Hurricanes Are Named

Hurricanes have been named since 1953. Each year before the hurricane season (June to November) the Tropical Prediction Center announces the names by which tropical storms and hurricanes will be known. When a tropical storm turns into a hurricane, it retains the same name. For several years only female names were used, until complaints were raised about sexism—and now the names alternate between male and female. There are none beginning with Q, X, Y, or Z because there are too few proper names that start with those letters. In 2005, the number of hurricanes for the first time exceeded the number of names reserved for them, forcing the Prediction Center to resort to its fallback plan, designating storms by letters of the Greek alphabet.

lift to the rising warm air, increasing wind speed even more.

The storm is kept organized by strong winds at higher altitudes (up to 30,000 feet or 9,000 m), which remove rising hot air from the storm's center. If the winds at higher altitudes were not at uniform speed, the storm would fall apart. A hurricane is composed of three parts: an eye, which is a low-pressure area, a calm center of the circulating system; an eye wall, dominated by the fastest, most violent winds of the storm; and rain bands, composed of thunderstorms circulating outward from the eye.

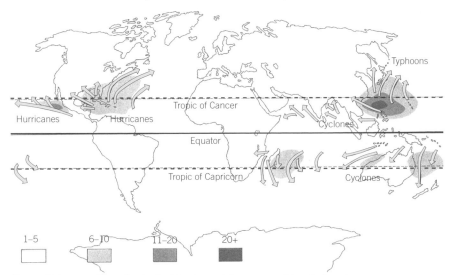

Above: *This map shows the typical tracks and frequency of tropical cyclones over a 20-year period.*

Floods and Droughts

Flooding has many causes, ranging from storms to runoffs from melting snow in spring; while it can bring vast destruction, flooding can sometimes be beneficial by leaving deposits of silt on the land, making it more fertile. Drought, on the other hand—defined as a protracted period without significant moisture—has nothing to recommend it.

Floods

Floods occur more often than any other natural disaster. Flood is defined as a large expanse of water that submerges land. Floods can occur as a result of a variety of factors, such as a heavy storm that coincides with a high tide, resulting in a storm surge; heavy and prolonged rainfall; and runoff from melting snow.

Land where there is not sufficient soil or vegetation to soak up excess water is prone to flooding. Deforestation can increase the chance of flooding and intensify its effect, as trees serve to anchor the soil. Floods that occur suddenly are known as flash floods. They usually occur after a concentrated rainfall in a small area and subside quickly. Unlike flash floods, river floods are slow to develop. They are often seasonal and are caused by long periods of heavy rainfall or the melting of deep snow over river basins. A river flood can remain for days or weeks.

Global warming is likely to increase the incidence of flooding, especially as polar glaciers begin to melt, releasing copious quantities of water into the oceans.

Not all flooding is bad: flooding can leave soil more fertile by depositing the land with silt, which explains why flood plains are so attractive for agricultural purposes. Without the annual flooding of the Nile, the ancient Egyptian civilization could not have flourished; ancient Mesopotamian culture, too, could not have prospered without the regular flooding of the Tigris and Euphrates.

Humans have been building dykes, dunes, levees, and dams for millennia as bulwarks against flooding. But these measures do not always work: coastal areas in the northeastern United States are constantly being swept away by winter storms in spite of multimillion dollar projects funded by the U.S. Government to restore the shoreline. Warnings can be

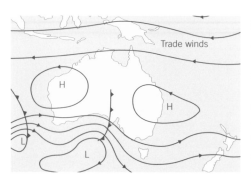
Trade winds

Left: *Australia is prone to recurrent periods of drought. The continent is located in the subtropical high-pressure belt, an area of sinking, dry air and usually clear skies. Some parts of the country—in particular the far north and south and the east coast—generally experience sufficient rainfall for most of the year, but rainfall over most of the country is both low and highly erratic.*

Monsoons

A monsoon is defined as a wind pattern that reverses direction on a seasonal basis. Monsoons are commonly associated with heavy, prolonged periods of rain and flooding. Monsoons occur in many parts of the world, but they are most powerfully felt in Asia. The monsoon season in southern Asia typically lasts from June to September, when winds blow in from the southwest. Monsoons in India develop when heat begins to rise off the land in summer, which creates an area of low pressure. It is this low-pressure system that creates the constant wind, which carries moisture from the oceans. Some of this moist air is diverted upward by the mountain ranges, notably the Himalayas; as it cools, condenses, and forms clouds, it produces heavy rainfall. Without the monsoons, the agricultural economy in India and her neighbors would grind to a halt. But floods produced by the monsoons can cause immense devastation, especially in Bangladesh, whose land mass is near or at sea level. The pattern reverses in the winter, when Australia warms in its summer sunshine while Asia cools during its winter.

Summer monsoon

Himalayas

Wet southwesterly winds

Indian Ocean

Winter monsoon

Himalayas

Dry northeasterly winds

Indian Ocean

given of approaching river flooding many hours or days before the event. Flash floods are harder to predict, but because radar can measure the amount of rain that is falling, it can be used to determine whether a flash flood is likely to develop.

Drought

Droughts are defined as a period of abnormally dry weather sufficiently prolonged for the lack of water to cause serious hydrologic imbalance in the affected area. A drought is not determined by the number of days, weeks, or months that go by without rainfall. It depends on the amount of available moisture before and after the period without rainfall and the length and size of the affected area.

A lack of rainfall is primarily responsible for droughts but a drought can also occur because the land itself has little ability to absorb or retain moisture. Climate change is likely to produce warmer winters, which will mean less snow and less runoff, resulting in more droughts throughout the world.

El Niño and La Niña

The periodic warming of the Pacific Ocean off Peru in early winter, known as El Niño, exerts remarkable influence over global climate, producing drier conditions in Southeast Asia and more rain in South America. La Niña, which, like El Niño, occurs every few years, develops from the cooling of the eastern tropical Pacific Ocean.

El Niño

El Niño is a periodic phenomenon produced by a weak, warm current in the Pacific off Ecuador and Peru. It is called El Niño (Spanish for "The Christ Child") because this pattern is usually seen around Christmastime. El Niño is a disruption of the ocean-atmospheric circulation system in the tropical Pacific. Its importance, however, extends far beyond the waters off South America; in fact, it affects weather worldwide with implications for the global economy.

El Niño is part of a larger, continual phenomenon known as El Niño-Southern Oscillation (ENSO), an irregular cycle of shifts in ocean and atmospheric conditions. El Niño occurs every three to seven years and can last months and occasionally up to a year. El Niño develops from the trade winds in the tropical Pacific. These winds drive the surface waters to the east; as these waters are exposed to the Sun they heat up and eventually begin to replace the cooler, nutrient-rich water off Peru and Ecuador.

Rainfall follows the warm water eastward and leaves the western Pacific with less precipitation. In this way El Niño is the cause of increased rainfall and flooding across the southern United States and Peru. It causes droughts in the western Pacific that make Australia vulnerable to devastating fires. The change in the surface temperature of the waters influences the atmosphere over the warmest water,

Upwelling

Upwelling is a normal oceanic process whereby deeper, colder water rises to shallower depths. This process brings nutrient-rich water from deeper levels to replace surface water that has dissipated by the action of winds, replenishing the food source for marine life. The surface layer is known as the mixed layer; a transition zone known as the thermocline separates the mixed layer from the deeper layer. As the depth decreases, so does the temperature. During an El Niño year, however, with warmer waters on the surface, the thermocline is deeper, limiting the upwelling process, which means that nutrients cannot rise to the surface as readily, depleting food sources for fish.

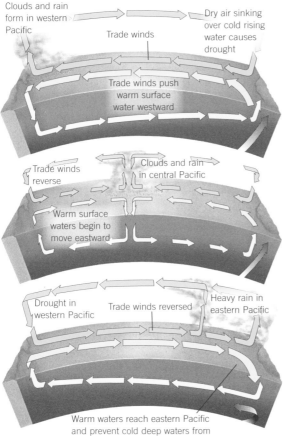

Clouds and rain form in western Pacific

Dry air sinking over cold rising water causes drought

Trade winds

Trade winds push warm surface water westward

Trade winds reverse

Clouds and rain in central Pacific

Warm surface waters begin to move eastward

Drought in western Pacific

Trade winds reversed

Heavy rain in eastern Pacific

Warm waters reach eastern Pacific and prevent cold deep waters from reaching the surface

La Niña

La Niña is a periodic cooling of waters in the eastern Pacific that causes temperatures to drop and brings drought to the region. The trade winds push the warmer waters westward, producing moist weather on the west side of the Pacific Ocean.

Transition

Every few years the conditions that created La Niña reverse as a result of a change in the direction of the trade winds, setting the stage for the formation of El Niño.

El Niño

El Niño produces warm temperatures and rain in the eastern Pacific when it forms in the waters off Peru; it also can bring warmer winters to North America while fostering drought in Australia.

causing changes in weather in regions much farther away. The impact of El Niño is most dramatic in winter, resulting in milder weather in the midwest and northeast United States, and cooler than normal weather in southern states.

Severe El Niño events have been responsible for thousands of deaths around the world, uprooted thousands from their homes, and caused billions of dollars of damage.

La Niña

La Niña ("The Little Girl") occurs when surface temperatures in the eastern

tropical Pacific become cooler than normal. The phenomenon is rarer than El Niño, occurring about half as often as her male counterpart. La Niña produces warmer winter temperatures than normal in the southeastern and southwestern United States, and below-normal temperatures in the northern United States—just the opposite effect from El Niño. Sometimes a strong El Niño can give way to strong La Niña conditions without any interruption, which is what happened in 1998 when ocean surface temperatures over the central Pacific dropped from about 1.8°F (1°C) above average to 1.8°F (1°C) below average.

Freak Weather

In the spring of 2006, California was struck twice by mysterious chunks of ice, one the size of a microwave oven, that plunged out of a cloudless sky. No one knows why. Some weird weather—aberrant natural phenomena—can be explained, but not all of it. And that is probably why it continues to hold such fascination.

St. Elmo's Fire

The large electric fields that build up between clouds and the Earth during thunderstorms are most intense at the tops of pointed objects. These fields may be so strong that they excite (or ionize) the air molecules. Sometimes when these excited molecules release energy it may be in the form of visible light. That's the phenomenon that mariners hundreds of years ago witnessed when they would see the top of their masts begin to glow. They called this glow St. Elmo's fire in the mistaken belief that the saint was protecting them from lightning strikes. What was really happening was that intense static electricity was coursing up the length of the mast and discharging a glowing corona. Schoolboys playing soccer in Dover, England, in 1976 observed a similar effect when they found that their heads were all aglow. Although St. Elmo's fire is not dangerous in itself, it can signal that a lightning strike is likely to occur shortly.

Ball Lightning

Another type of light show associated with thunderstorms is ball lightning. This electrical freak of nature is still a subject for speculation. In June 1996, for example, workers at a printing factory in Tewkesbury, England, were astonished to see a sphere of blue and white light

Strange Weather

Of all the weird weather reported, thunderstorms and rainfalls can produce some of the most freakish. Dead birds have been known to fall out of the sky, most likely sucked up by updrafts in a thundercloud (where they froze) or from tornadoes. Large chunks of ice have also descended to Earth, most recently in California in 2006. While some scientists have reasoned that the ice has fallen off passing aircraft, there is little evidence to support their view. A chunk of ice 20 feet (8 m) in length fell in Scotland in 1849, half a century before the invention of airplanes. There have been instances where people have reported blood falling from the sky, but the "blood" turns out to be red dust picked up by the wind in the Sahara Desert and carried thousands of miles inside a high-pressure system before falling back to the ground in a rainstorm.

the size of a tennis ball spinning along the building's girders and hurtling into machinery, producing a flurry of sparks. Then the mysterious ball of light struck a window and exploded with an orange flash, knocking out the telephone switchboard. Three people sustained electric shocks during the surprise visitation, none seriously. This event was by no means unique; thousands of people have witnessed such mysterious balls of light.

Ball lightning comes in a variety of sizes; some have been reported as large as a football while others are no larger than a golf ball. They never make any noise or produce any odor. Almost invariably they vanish with a pop when they hit an appliance like a television. There are, however, a handful of cases where they explode, as the ball did in Tewkesbury, erupting into flames and setting a house on fire.

Waterspouts

A waterspout is essentially a small tornado over water that usually takes form as a funnel-shaped cloud. There are two types of waterspouts: weaker "fair-weather waterspouts" and much stronger "tornadic waterspouts." Waterspouts generally appear in association with a cumulus-type cloud (see pp. 78–9). They have a similar basic structure to tornadoes, typified by upward-moving air. The funnel cloud owes its origin to the action of rushing winds at the water surface; as the winds intensify they swirl into a vortex and begin to rise up. When the wind speeds reach around 40 mph (65 km/h), the wind causes spray to form a circular pattern—the spray vortex. The funnel cloud that emerges from this process extends all the way from the ocean to the parent cumulus cloud, but the vortex can only be seen at an altitude high

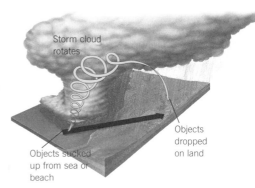

Above: *Waterspouts are tornadoes that form over water, typically in regions where thunderstorms are frequent. These funnel clouds originate from cumulus clouds.*

enough for the lower pressure to condense the water vapor into water droplets. (That is why it looks as if waterspouts are not touching the water.) When the spray vortex weakens, the funnel shortens and tapers, ultimately dissipating.

Waterspouts can attain speeds of up to 190 mph (300 km/h). There are theories that waterspouts account for some of the disappearances of vessels in the Bermuda Triangle. Although most waterspouts occur in the tropics, they have been known to appear in temperate areas such as the western coasts of Europe. They have also occurred on freshwater.

Whirlwinds

Whirlwinds are formed in much the same way as tornadoes. They are small, swirling columns of air and are often called dust devils. A whirlwind usually occurs over deserts and semiarid plains on hot, calm days when the heated surface of the Earth produces a mass of overheated air just above the ground. This air mass rises in a column and begins to suck up debris.

Weather Forecasting

Humans have been trying to predict the weather for millennia, with mixed results. The science of meteorology began with the development of instruments such as the thermometer and barometer in the 17th and 18th centuries. Today, with the help of advanced technology, such as satellites and supercomputers, forecasts are becoming increasingly accurate and detailed. Perfection, however, will probably never happen.

You may have heard of the butterfly effect, the idea that a small event in one location can produce large events on the other side of the world, days or weeks after the fact. That is the way it is with weather, too: a small change in atmospheric conditions in one place may have a dramatic impact halfway across the world five days later. A situation fraught with so much unpredictability poses quite a challenge for meteorologists. It explains why weather forecasters seldom predict the weather more than five or six days ahead. Longer-range forecasts tend to be more general, based on average climatic conditions.

Before the TV weather person goes on the air to tell you about tomorrow's outlook, national weather services and private weather-forecasting companies have to collect meteorological data from all over the world. Once processed by supercomputers using mathematical models of the atmosphere, the data emerges in the form of predictions.

Forecasts are not only needed by the general public. Forecasting information is vital to the economy: for example, it is estimated that as much as one-third of the U.S. gross domestic product is at least partially dependent on weather.

History of Forecasting

People have probably been trying to forecast the weather since they started planting, as their very survival depended on it. And while these forecasts could not be considered "scientific," careful observation over the millennia produced a body of practical knowledge. Much of what we think of as weather folklore grew out of these observations and proved surprisingly useful to our forebears (see pp. 106–9).

Since water in its vapor state is invisible, its existence was not understood until the Renaissance. Without an understanding of water vapor it was impossible to understand how weather works. In 1450, Cardinal Nicholas de Cusa is credited with an examination of the properties of water vapor, which he called "invisible water." The science of forecasting—meteorology—can be traced to the mid-17th century with the invention of the barometer, ascribed to the Italian Evangelista Torricelli in 1643. For the first time an instrument was available to measure changes in atmospheric pressure, which made it possible to predict approaching storms with reasonable accuracy. In 1664, with the invention of the hygrometer by Francesco Folli of Italy, it became possible to measure humidity. (A very basic humidity-measuring device was invented in the mid-15th century by Cardinal de Cusa.)

The thermometer was added to the meteorologist's toolbox in 1714 when the German Daniel Fahrenheit developed the mercury thermometer. (The invention of the original, water-based, thermometer has been attributed to Galileo Galilei, possibly in 1593.) By the late 18th century, the French scientist Antoine Laurent Lavoisier was able to take daily measurements of air pressure, moisture content, wind speed and direction. "With all of this information," he declared, "it is almost always possible to predict the weather one or two days ahead."

Sometimes advances came about through disasters, as in 1854 when a French warship and 38 merchant ships sank in a storm off Crimea. An investigation launched by the Paris Observatory found that the storm had developed two days before and swept across Europe. If the storm had been tracked, the calamity might have been avoided. This led to the establishment of the first national storm-warning service in France in 1863, the earliest of its kind.

Communication remained a problem, as it was difficult to get observations made in one location to another to warn of an approaching storm. The problem was solved once the electric telegraph, invented by Samuel Morse in 1831, came into widespread use. With increasing use of the telegraph, the Paris Observatory was also able to obtain more detailed information about weather from many more locations on a timely basis. The next step was to disseminate this information, and the first published weather maps were the result.

Aristotle

By most accounts we owe the term "meteorology" to the famous Greek philosopher Aristotle, who wrote a book called *Meteorologica* in 340 B.C. The book is a compilation of everything that was known about weather at the time, including observations of the atmosphere, clouds and mist, rain, snow, wind, hail, lightning and thunder, and climatic changes. Never one to shy away from ambitious projects, Aristotle also tackled astronomy, geography, and chemistry in his work. The application of the word meteorology to weather is due to the ancient Greek practice of naming anything that fell from the sky—rain, snow, ice, or planetary debris—a meteor. Contemporary meteorologists refer to precipitation as "hydrometeors" (particles of water or ice in the atmosphere) as opposed to extraterrestrial meteoroids. Although Aristotle was mistaken in many significant respects—he tried to analyze atmospheric phenomena using philosophical concepts—it was not until the 17th century that the science had advanced sufficiently to show that he was wrong.

Modern Forecasting

As a science, meteorology has made significant strides in the last century due to new technology and new mathematical models intended to simulate weather conditions in the real world. Today forecasters rely more than ever before on weather satellites, radar, weather balloons, and observations from automatic weather stations to collect data that is then processed by supercomputers.

Predicting the Future

The 20th century saw a quantum leap in technological progress with the development of modern weather stations, radar, cameras, weather balloons, satellites, and supercomputers. But technology alone would not be of much help without the theoretical underpinnings to apply it. Shortly after World War I, for instance, British meteorologist Lewis Richardson determined that the atmosphere was governed by the laws of physics, which led to the development of mathematical models as a predictive tool. Even so, his formula was difficult to use for calculations because it was extremely complicated, a problem that would only be addressed by the invention of computers that could process huge quantities of data in the blink of an eye.

Today one computer can analyze data from the more than 3,500 observation stations around the world and derive a forecast of what the world's weather will be for the next 15 minutes. But because weather systems are so complex and depend on so many factors, no computer, no matter how powerful, can produce a forecast with 100 percent accuracy. Still, modern meteorologists do a fairly good job. Britain's Meteorological Office, for example, claims an 86 percent accuracy for its 24-hour forecasts and 80 percent accuracy for its five-day forecasts.

In making their forecasts, meteorologists focus on three scales: synoptic, mesoscale, and vertical. The synoptic scale covers weather on the broadest level: the movement of air masses, fronts, and pressure systems. The mesoscale is the local scale and examines the effects on weather of topography, bodies of water, or heat produced by cities in a particular geographical location. The vertical scale looks at the structure of the atmosphere: how pressure, temperature, and density alter with altitude.

Meteorologists can choose from, and combine, several different methods to make their forecasts. There is, for instance, the persistence method, which is predicting the weather based on the pattern of weather in the past (tomorrow will be much like today). This method is limited to regions where weather is relatively stable (the southeastern United States in the summers, for example). The climatology method relies on averages of temperature, precipitation, humidity, and so on over many years. This method, too, is obviously limited since weather often does not conform to precedent. Numerical Weather Prediction (NWP), which relies on supercomputers to do a staggering amount of number-crunching based on observational data fed into them to produce forecasting models, is preferred because it can produce much

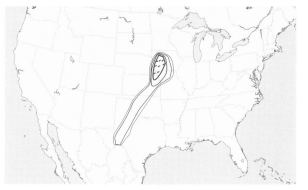

Left: *Many forecasts today are given in the form of probabilities. This map depicts the probability of tornadoes in the United States over a given period. The lines indicate different probabilities, rising from 2 percent in the outer ring to 15 percent in the inner.*

more accurate forecasts. However, NWP, too, is limited because the equations used by the models to simulate atmospheric conditions cannot be entirely precise.

Clues from the Past

Meteorologists today are not only relying on computer models to predict the future: they are also looking to the past for clues about the future. While most scientists agree that important climate changes are under way, they remain uncertain about what kind of changes. That is where history—very ancient history—can offer guidance, especially where the build-up of greenhouse gases is concerned. In 2006, scientists published findings showing that 55 million years ago the Arctic Ocean was much warmer than had previously been imagined, with an average temperature of 74°F (23°C). They based their findings on fossils of ancient organisms and rock samples at a location not far from the North Pole. The fossils came from an epoch of global warming known as the Paleocene Eocene Thermal Maximum. The scientists believe that that warming was caused by an accumulation of heat-trapping greenhouse gases such as methane and carbon dioxide in the atmosphere—with the same effect that is being seen today.

However, at that time, the build-up of gases was a result of natural causes—no humans or human precursors were on the planet then—but so far scientists have been unable to pinpoint them.

The Arctic samples also depict the subsequent history of climate change: about 45 million years ago a long period of cooling began, culminating in a cycle of ice ages alternating with intervals of warming. An examination of the fossils indicated that the cooling trend was triggered by a significant decrease in the concentration of greenhouse gases. These findings suggest that climatologists will have to rewrite the history of climatic change on Earth. At the same time they have crucial implications for our future. That climate can change so quickly as a result of the increase of these gases in the atmosphere and bring Florida-like warmth to the Arctic means that we may see similarly dramatic changes within a relatively short time. Change might not occur incrementally over a period of hundreds of thousands or millions of years, as scientists once expected, but rather might occur within decades. "Something extra happens when you push the world into a warmer world, and we just don't understand what it is," said Henk Brinkhuis, one of the researchers involved in the Arctic study.

Professional Weather Stations

A weather station monitors atmospheric conditions and collects meteorological data using an array of instruments and sensors. The data acquired from thousands of local weather stations are used to produce the weather forecasts issued by meteorological agencies.

A weather station serves as a location that monitors atmospheric conditions and collects meteorological data using an array of instruments and sensors. Meteorological agencies use the data acquired from weather stations to produce weather forecasts. In the United States this network of weather stations is administered by the National Weather Service, one of six scientific agencies run by the National Oceanic and Atmospheric Administration (NOAA). In Australia, professional weather stations operate under the authority of the Bureau of Meteorology, in the U.K. it is the Met Office, in Canada the Meteorological Service of Canada, and in China the China Meteorological Administration. In addition, many universities and private companies process and analyze meteorological data collected from weather stations for clients with specific needs, focusing on weather that might affect farmers or maritime interests, for example.

There are no formal international standards governing the type or quality of equipment used by weather stations. But because meteorological data circulates around the world, some conventions are necessary. Weather maps use internationally agreed-upon symbols, and meteorological reports are issued using Universal Time (UT). In addition,

several informal international organizations share data and address common problems—for example, the World Meteorological Organization (WMO), an agency of the United Nations, and the Coordination Group for Meteorological Satellites (CGMS).

Weather stations around the world report the current temperature, maximum and minimum temperatures for the day, humidity, atmospheric pressure, wind speed and direction, cloud cover, and collect radar images. The intensity of precipitation can be predicted for about 100 miles (170 km) around the radar. No weather station would be complete without certain essential instruments:
• Thermometer to measure temperature
• Barometer to measure barometric pressure
• Hygrometer to measure humidity
• Anemometer to measure wind speed and direction
• Rain gauge to measure precipitation
• Psychrometer to measure the strength of evaporative cooling in the atmosphere
• Ceilometer to measure cloud height

In addition, stations in areas that experience regular deep snowfalls will employ a device called a snow stake to measure the depth of snowfall. A snowboard, which is simply a base for snow to fall on, is also a common feature in most weather stations.

Manual and Automatic Weather Stations

Manual weather stations are the traditional type and require people to run them and monitor weather conditions. By contrast, automated weather stations rely on sensors to measure and record weather conditions every 30 secs. The most advanced automated system is the Automated Weather Sensor System (AWSS), which replaced the Automated Surface Observation Systems (ASOS). These systems record temperature, precipitation types and accumulation, wind direction and speed, humidity and dew point, barometric pressure and visibility, sky cover and ceiling as well as thunder—all of crucial interest to pilots. These units can measure freezing rain by means of a vibrating wire that stops moving as soon as enough ice accumulates. These weather stations are also capable of sending out warnings via VHF radio signals when weather conditions are undergoing a significant change.

Taking Readings

The location of a weather station is an important consideration in the analysis of its data—it might be near a body of water, for instance. Without knowing where the station is situated, it is difficult to determine the relevance of its data to a location at a considerable remove. The location is defined as the point (or points) at which the measurements are taken. For instance, the exact altitude is fixed by international convention based on measurements of different types of barometers or, more simply, the base of the rain gauge. Topographic maps are employed to determine altitude and location of the site to the nearest foot and minute of degree, respectively.

Weather stations maintain records based on what are called "spot" observations, or observations taken from instruments at certain times of the day, every 30 seconds in the case of the most modern automated stations (see "Manual and Automatic Stations," above).

In contrast to weather observations, climatological records are based on observations only taken at the end of the day, preferably at midnight, when climatic variables are recorded, including highest temperature, lowest temperature, rainfall, snowfall, and prevailing winds. Climatological records are used to measure long-term weather conditions at the station.

To aid observations, weather stations also maintain visibility charts based on topographical maps and photos that indicate the location of observable buildings, as well as natural objects such as hills and lakes. Nighttime charts indicate the visibility of such features as green or red airway beacons, red TV and radio tower obstruction lights, red collision lights on buildings, and street lights.

Environmental Satellites

Weather satellites provide meteorologists with an unparalleled view of the Earth's surface from space. They provide invaluable information about the world's weather—particularly weather events that occur in remote locations where there are no witnesses on the ground.

Since the launching of the first environmental satellite, Vanguard 2, by NASA in 1959, satellites have provided meteorologists and scientists with invaluable information about the state of the Earth. These satellites monitor pollution, city lights, volcanic activity, dust storms, forest fires, ocean currents, clouds, and snow cover. Satellites also track ice floes, an increasingly vital mission due to the problem of global warming.

Information Collected

At a basic level, virtually no expertise is necessary to interpret visual images produced by weather satellites. It is easy to make out clouds, fronts, tropical storms, and snow-covered landscapes. It is even possible to discern the direction and strength of the wind by following the patterns of clouds, their alignment and movement illustrated by successive photos.

Satellites are equipped with a variety of sensors. Radiosensor scanners, for example, capture thermal or infrared images. These images allow analysts to determine cloud heights and types, and land- and water-surface temperatures. Surface temperatures are critical for farmers and fishermen, in the former case because it helps protect crops from frost, in the latter because it helps identify where stocks of fish are likely to be

Weather Balloons

Weather balloons are used to track temperature, pressure, relative humidity, and wind speed and direction in the upper atmosphere, monitoring regions that are often inaccessible to other types of technology, including weather satellites. About 2,000 balloons are launched around the world every day. Typical balloons are approximately 6 feet (2 m) in diameter and lifted aloft by helium or hydrogen; they contain a package of instruments called a radiosonde. Flights usually last for about two hours at a maximum altitude of 22 miles (35 km). The radiosonde sends a constant stream of data to the weather service. If it is exposed to temperatures that fall below -130°F (-90°C) the balloon's diameter swells and the balloon pops because of low air pressure. The package of instruments is assured of a safe trip to the ground by means of small parachutes. The package also contains mailing instructions so it can be returned to the weather service.

Radar

For forecasters there are few more useful tools than radar. Weather radars send out radio waves from an antenna. Those radio waves are either scattered or reflected back to the antenna by objects they encounter in the air. These objects are often meteorological phenomena—raindrops or hailstones—but they can also be insects and dust. The radar electronically converts the radio waves in such a way that forecasters can see pictures that indicate the location and intensity of precipitation as well as wind direction and velocity. The most widely used type of radar is Doppler, named for its inventor, Christian Doppler, who used sound waves to develop his concept in 1842. The Doppler radar relies on changes in frequency as the radio waves are reflected back. Objects moving away from the antenna will change to a lower frequency and those moving toward it will change to a higher one. These changes offer a means of determining wind speed and direction as it blows around precipitation, insects, or other objects.

more common. Infrared also permits researchers to see into the depths of the sea to track the flow of such currents as the Gulf Stream. Climatologists can use the data to study the effect of climate change on the oceans.

Thanks to satellites, we now know that dust storms, once associated only with deserts, have a great deal of influence over global weather patterns. Dust storms forming over the Sahara drift across the equatorial region of the Atlantic Ocean, where they suppress the formation of hurricanes. Environmental satellites allow researchers to pinpoint the location of fires in remote areas. Satellites measure the waste of energy during the burn-off of gas in the oil fields of the Middle East and Africa.

Types of Satellites
There are two basic types of weather satellites: geostationary and polar orbiting.

Geostationary satellites orbit the Earth above the equator at altitudes of 22,300 miles (35,880 km). They orbit at the same rate that the Earth spins, meaning that they can monitor the same region 24 hours a day.

Polar orbiting satellites pass over the north and south poles with each revolution (Sun-synchronous orbits). As the Earth rotates to the east, the satellite monitors an area to the west of its previous pass. They view every location beneath them twice a day. Polar orbiting satellites are needed over the poles because the equatorial position of the geostationary satellites does not permit them to clearly see these regions. In addition, polar satellites occupy much lower orbits, about 440–500 miles (720–800 km), which means that they can provide far more detailed information about the polar regions.

Supercomputers

Supercomputers capable of processing billions and even trillions of bits of information every second are increasingly being used by national weather services to generate weather forecasts and track climate change. Using complex mathematical models to simulate weather patterns under various conditions, supercomputers represent the cutting edge of forecasting.

Forecasting Machines

"Thirty years ago people were delighted when we occasionally got a forecast correct," said James Hoke, director of the Hydrometeorological Prediction Center of the U.S. National Oceanic and Atmospheric Administration (NOAA). "Now their expectations have risen." And one of the chief reasons those expectations have soared is because of the role of supercomputers, which are increasingly being mobilized to analyze data and generate increasingly accurate forecasts. The NOAA, the umbrella agency of the U.S. National Weather Service, maintains three IBM supercomputers: a primary forecasting computer (Blue), a research computer (Red), and a back-up (White). Together, Red, White, and Blue form a major component of the Global Earth Observation System of Systems (GEOSS).

Of course, no computer, whatever its size, is of any use without the right kinds of data being fed into it. When it comes to the weather, that data streams in around the clock from a staggering variety of sources: mountaintop observation stations, environmental satellites, ocean buoys, radar, and some 700 global weather balloons. In addition, wide-body aircraft carry sampling instruments that monitor atmospheric conditions. Of all these sources, however, satellites are providing the most extensive coverage. The number of daily weather observations fed into NOAA's supercomputers amount to over 200 million a day. In turn, the supercomputers turn out some 200,000 forecasts, warnings, and other types of information or "products" every day. Products include maps of temperature ranges or wind patterns, and complicated regional and global meteorological models.

One result of all this number-crunching is that forecasts are becoming ever more reliable. "The four-day forecast is now as accurate as the old two-day forecast," one meteorologist noted. "The two-day forecast is as accurate as the old-one day—though after about 72 hours, it's still difficult to get an accurate forecast from these computer models." These models are then used by meteorologists to predict long-range weather and climatic patterns—assessing the potential intensity of hurricane seasons before they begin in June, for example.

Supercomputers also provide researchers with simulations of weather and climate. The challenges are formidable, however: just to make one climate simulation, a 100-year climate model might have to run for hundreds of hours on a very large supercomputer. As a result, hundreds of such runs must be made in order to test and improve a climate model. The objective is to provide scientists and the public with a much clearer picture of what can be expected from climate in the coming years.

What is a Supercomputer?
A supercomputer is a computer with an extraordinarily high capacity for processing calculations. It derives its name from a 1920 article in the *New York World*, which used the term "super computing" to describe an IBM tabulator made for Columbia University. However, the first supercomputers were not introduced until 1960, manufactured by Seymour Cray of the Control Data Corporation (CDC). The Cray machines, later produced by Cray's own firm, dominated the market until the early 1980s, when competitors got into the game.

We have come a long way from the early computers of the 1950s that worked at speeds of about a thousand calculations per second, a rate known as a kiloflop. Today supercomputers are processing data at a rate of trillions of operations per second—a speed measured in teraflops. Each supercomputer run by NOAA can run at more than 4 teraflops per second, but when working together, can perform 7.3 trillion calculations per second, or 7.3 tetraflops. But computer developers have set their sights even higher, contemplating speeds measured in the petaflops. That's the equivalent of processing data contained in a stack of paper 60,000 miles (100,000 km) high every second! Put another way, they will be performing more than 1,000 trillion calculations per second. As processing speeds increase, so should the accuracy of the forecasts. When NOAA's supercomputers, for instance, went online in 2000, they were 28 times faster than the Cray supercomputers they replaced, which were installed just six years previously. That accounted for an estimated 10 percent increase in accuracy of forecasting temperatures and humidity and in pinpointing when, where, and how much rainfall will occur.

In the U.K., the Meteorological Office installed two NEC SX-6s supercomputers in 2004. It was a fitting way to celebrate the Met Office's 150th anniversary. It was hoped that the new machines could cut in half the number of "busts"—those occasions when forecasters fail to get it right. As powerful as they are, though, at the time of their installation they were still ranked as only the 275th and 276th in the world, with capacity of slightly less than one tetraflop per second.

In some ways, supercomputers have lost much of their allure as smaller computers acquire greater power. Now they are mostly custom-made for clients by such companies as IBM and Hewlett Packard. Not everyone is enamored with the potential of supercomputers. Critics cite advances in technology and data-storage systems to argue that these high-powered systems are becoming obsolete and that distributed computing systems—in which computer-processing capacity is shared by many less powerful computers working in tandem—are more effective. All the same, the detractors still seem to be in the minority.

Reading Weather Maps

All those bars, arrows, and barbed lines you see in a weather map have a specific meaning. These symbols can provide a visual picture of developing weather patterns over a particular locale or an entire continent. There are many types of weather maps: some, for instance, focus on temperature, others on wind conditions, still others on precipitation.

To some degree, everyone is familiar with weather maps from television and the newspapers. But how to go about interpreting these maps and make sense of all those arrows and blobs may not be quite so clear. Weather maps are by no means comprehensive: there's a lot more going on in the atmosphere than these maps could possibly show and still make

any sense. Rather, a weather map is a simple representation of surface weather systems—highs, lows, fronts—that have already occurred or are likely to occur in the near future.

There are different types of maps focusing on particular features, such as surface maps, dew-point maps, temperature maps, and pressure maps.

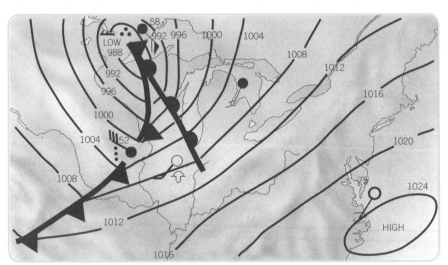

Above: *High-pressure systems are often represented by H and low-pressure systems by L. Isobars, which look like the contour lines on a terrain map, provide detailed information about air pressure and wind speed and direction. The isobars will be closer together in low-pressure systems (indicating cloud cover, precipitation, strong winds, and potentially stormy weather) and farther apart in high-pressure systems (indicating clear skies, dry conditions, and very light winds). Generally, winds blow along the line of the isobars, but they turn slightly toward a low and away from a high. The closer together the isobars, the stronger the winds. Fronts are represented by lines defining the boundary of the front. Cold fronts feature triangles along the length of the line while warm fronts feature hemispheres.*

Weather Map Symbols

Meteorologists around the world use the same symbols so that they can exchange information easily. The meaning of some symbols is not obvious to the public, so the symbols used on television and newspaper weather maps tend to be simpler. Meteorologists measure windspeed in knots: 1 knot is 1 mph (1.85 km/h).

light drizzle	freezing rain	no clouds
steady, heavy rain	tornado	partially overcast
light snow	dust or sand	completely overcast
steady, light snow	fog	stratus clouds
hail	lightning	cumulus clouds
	hurricane	windspeed 10 knots
		windspeed 105 knots

Showing Pressure and Fronts

The most prominent feature on all these maps is the pattern of high and low pressure. The former are often marked with an H, and the latter with an L. The weather map symbol for a cold front is a blue line with triangles pointing the direction the cold air is moving. The surface location of a warm front is marked with a red line of half circles pointing in the direction of travel. Shaded areas can show where rain has occurred in the previous 24 hours.

Showing Wind Speed

Wind strength and direction can be read by studying the pattern of isobars. Isobars are the lines that surround areas of high and low pressure, connecting areas of equal barometric pressure. Isobars are measured in millibars, a unit of atmospheric pressure (see pp. 14–15). Mean atmospheric pressure at sea level is approximately 1013 millibars. These isobars serve as markers that can indicate wind speed. The closer these isobars are to each other, the stronger the pressure gradient—that's the amount of pressure change over a given distance. The stronger the pressure gradient, the stronger the wind will be. (This is not true in the tropics, however, because the force of the Earth's rotation is weak.) Wind direction is shown by arrows with barbs on their tails to indicate speed.

MAKING WEATHER PREDICTIONS

In spite of all the washed-out barbecues, weather forecasting is much more accurate than people generally believe. Professional meteorologists are right about 80 percent of the time. Though professional meteorologists may have the advantage of supercomputers, the Internet has leveled the playing field. Amateurs now have access to real-time weather data, including radar and satellite imagery, streaming in from all over the world. And while the technology used by national weather stations is likely to be more advanced than anything you can buy, professionals still rely on the same essential instruments that anyone can acquire at relatively low cost. Moreover, many software packages are available for even the most budget-conscious weather enthusiast.

In the chapters that follow, you will learn how to set up your own weather station and then monitor it effectively, using its data to make your own forecasts. You will also learn how to analyze cloud cover to assess present and future weather conditions, and get some extra tips on such key factors as how to predict different forms of precipitation, and what effects pressure systems and winds will have on your weather. Armed with all this information, you may even beat the professionals at their own game from time to time. Amateurs can share their observations of local conditions with other enthusiasts online, and a growing collaboration between professionals and amateurs is leading to more accurate and detailed forecasts.

Setting Up Your Own Weather Station

Anyone can be a meteorologist. It does not require a lot of money or training to make observations or even to produce forecasts. What it does require is a simple weather station. Weather stations can be built or purchased ready-made. They consist of several basic instruments, such as a thermometer and anemometer, and a shelter to house them.

For those who do not want to build their own station it is possible to purchase one that uses Windows software for downloading and viewing data on a PC. These stations come with sensors to measure wind speed, wind direction, temperature, humidity, atmospheric pressure, and rainfall, and can operate using a regular television antenna (see pp. 68–9).

You Will Need

Shelter (Stevenson screen)
Flashlight or **low-lantern lighting**
Clock (Universal Coordinated Time or Local Standard Time)
Visibility chart
Thermometer (maximum-minimum, precision, or digital)
Barometer (mercury, aneroid, or digital)
Anemometer
Hygrometer or **psychrometer**
Rain gauge (8-in gauge, 4-in gauge, or wedge gauge)
Snow board or **snow can**
Weather journal

Positioning Your Shelter

If you are not buying an electronic station, first of all you will need a shelter to house your instruments to protect them from exposure to the elements, which not only carries the risk of dirt and moisture but can distort the readings they give. The shelter also should provide adequate ventilation. Weatherproof boxes are usually made out of wood or plastic. The box should be painted white to prevent it from absorbing heat. The shelter should have a double top, with louvered walls with slats sloping downward from the inside to the outside of the shelter. The thermometer should be attached to the bottom of the box. Government-standard shelters can be purchased or built to specification.

Once you have bought or built the shelter, your next step is to find a safe location on whatever side of your house gets the most shade. Choose a spot in the shade in a grassy area, at least 100 feet (30 m) away from paved or concrete surfaces, which absorb heat. You should not place it

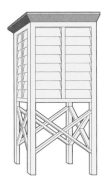

Left: *A Stevenson screen is the most common shelter to house most instruments necessary for a functioning weather station.*

in either a hollow or on top of a hill. The shelter should not be any closer to an air-conditioned building than four times the height of that building. That means that a shelter should be positioned at least 60 feet (18 m) from a 15 feet (4.5 m) structure.

Shelters must be elevated because the thermometer should be about 5 feet (1.2 m) above the ground. The shelter should be as secure as possible to keep instruments from being jarred. The door should open in a direction to avoid sunlight. To see inside the shelter at night you should either use a flashlight or install a low-power lantern bulb. Make sure that you do not use matches, cigarette lighters, or standard lightbulbs. You should avoid breathing on the instruments because the condensation from your breath can affect accuracy.

Once you have established your site you need to determine your precise altitude and location to the nearest foot or meter and minute of degree respectively. Topographic maps available from government agencies in many countries, or GPS units, are recommended to plot your location.

Locating Instruments

The thermometer needs to be exposed to free-flowing air, which is why it is so important for the box to be vented. You might also wish to measure the soil temperature to determine how warm the ground is, how much heat might radiate into transient air masses (which can affect their stability), and how fast ice or snow will melt once it has fallen. To obtain soil temperature you can route the thermometer's probe through a conduit into the soil. The "outdoor" bulb of the thermometer will give you the temperature of the soil. In the absence of a shelter, the thermometer should be mounted on the north side (south side in the southern hemisphere) of a thick white post in a grassy area away from buildings, but take care to mount the thermometer so that it does not absorb heat from the post itself.

The hygrometer (to measure moisture in the air) should be placed in the shelter.

The barometer should be kept inside your house, preferably on a heavy table and away from air-conditioning vents.

The anemometer should be placed clear of buildings and never installed on rooftops, because buildings will create wind shear problems. If there is no alternative to a rooftop you should place the anemometer at least 10 feet (3 m) above the roof. The U.K. Met Office recommends that they be installed at a distance from any building or obstruction of twice the building's height.

The rain gauge should be kept at a distance of at least 10 feet (3 m) from the shelter and away from all buildings and trees to prevent wind-driven rain from seeping in. Ideally, a gauge should be placed in a grassy area and surrounded by gravel or pebbles to protect it. A wind screen, placed at about 1 foot (30 cm) from the gauge, will help prevent loss of precipitation. See also pp. 66–7.

The snow board or snow can should be placed with the rain gauge.

Thermometer

Thermometers usually measure temperature on two different scales: Fahrenheit and Celsius. There are several types of thermometers, including precision and digital thermometers, but the most common is the bulb thermometer with a narrow tube containing mercury.

Temperature Scales

We owe the Fahrenheit scale to Daniel Fahrenheit (1686–1736), who decided that the freezing and boiling points of water would be separated by 180°, which meant that the temperature at which water freezes is 32°F and the temperature at which it boils is 212°F. The Celsius scale was invented by Anders Celsius (1701–44), who put the freezing point of water at 0°C and the boiling point at 100°C.

Constructing a Thermometer

The simplest type of thermometer is the bulb thermometer, which is made of a narrow glass tube and is composed of a liquid, usually alcohol. This type of thermometer is based on the principle that liquid will change its volume—expanding or diminishing—relative to its temperature. Liquids will take up less volume at cooler temperatures and more volume when temperatures warm. (The same principle applies to gases.) The reason that the tube is narrow is to emphasize the change in volume. To understand the principles behind a

thermometer, try constructing your own. You will need:

1 A glass jar or bottle with a screw-on watertight lid
2 A drill or a hammer and a large nail
3 Silly putty, plumbers putty, or caulk
4 A thin drinking straw
5 Food coloring

Drill a hole in the lid of the jar the right size for the straw to fit snugly inside it. Insert one end of the straw into the hole and seal the hole with the putty or caulk on the inside and outside of the lid. Then fill the jar to the very brim with water chilled by ice or kept overnight in the refrigerator. Now add the food coloring and shake. Screw on the lid. Now put your thermometer into the sink. Plug the sink and run hot water until it is about half full. You'll see the water expand inside the straw. If a Fahrenheit or Celsius scale were calibrated along the length of the straw you would be able to determine the temperature with reasonable accuracy. This homemade thermometer does have some

Temperature Conversions

To convert from Celsius to Fahrenheit:
([9/5] x C) + 32 where C is the temperature in Celsius

To convert from Fahrenheit to Celsius:
(F - 32) x (5/9) where F is the temperature in Fahrenheit

Keeping a Weather Log

Do not rely on your memory alone for monitoring your weather station. Keep a weather log like the one provided with this book. You should note the minimum and maximum temperatures every day, wind speed and direction, sky conditions, precipitation and amount of accumulation, and visibility. Many experts recommend taking instrument readings once in the morning and once in the evening, always at the same times. When instruments are automated you have the option of logging in data at much more frequent intervals. Measurements should always be reported in appropriate and consistent units. All data should be posted in Universal Time (UT). Many software programs will make the adjustment from local time to UT automatically. You should also write a brief summary of current weather conditions, including any information about fronts or storm activity influencing the weather where you live.

In some cases you will need to use two or more instruments in tandem. For example, forecasting future storm activity is best achieved through the use of a barometer together with a rain gauge and wind direction sensor. If pressure readings drop and wind velocity rises, you are probably due for a thunderstorm. You should also learn how to interpret readings. For example, before you can understand the meaning of a barometric reading you should first check the weather history for your locale to find the average barometric pressure so that you will know how to place a reading in a larger context.

important limitations. Below freezing the temperature cannot be measured on it because the water would turn to ice. Moreover, the size of the jar slows down the time it takes to register a temperature, by as much as an hour. For these reasons, most thermometers use alcohol rather than water, in a narrow glass tube.

Thermometer Types

A **maximum-minimum thermometer** is shaped like the letter "U." The mercury resides in the lower half of the "U" and moves through the tube to register changes in temperature by pushing up a metal index marker enclosed within the tube on each side. It is accurate to about one degree. The index marker remains at the maximum and minimum temperatures reached on each occasion.

Precision electronic thermometers, which are more expensive, are preferred by many weather agencies due to their high levels of accuracy.

Digital thermometers use microprocessor technology and the thermistor, a small device which changes its electrical resistance as the temperature changes.

Hygrometer

The amount of moisture in the air at a particular location—the level of humidity—is an important indicator of future weather. Hygrometers or psychrometers are useful devices for measuring humidity.

The hygrometer is one of the oldest weather instruments. An early version was invented by Leonardo da Vinci in the 1400s. A more practical hygrometer was devised in 1664 by Francesco Folli, but the modern dew-point hygrometer owes its invention to British chemist and meteorologist John Frederic Daniell in 1820.

The device is intended to measure absolute humidity—the temperature at which dew or frost will form—in order to determine relative humidity. The temperature at which dew forms is called the dew point. Relative humidity is a measure of the percentage of moisture in the air compared to what the air is actually capable of holding at a particular temperature. Absolute humidity is defined as the amount of water vapor that can be held in a specific volume of the air. Clouds will form and rain will fall in regions of the atmosphere where the air is at 100 percent

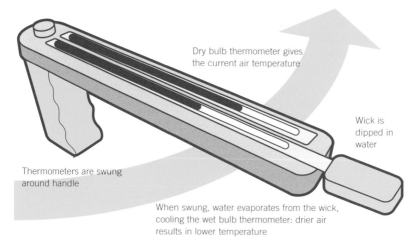

Dry bulb thermometer gives the current air temperature

Wick is dipped in water

Thermometers are swung around handle

When swung, water evaporates from the wick, cooling the wet bulb thermometer: drier air results in lower temperature

Above: *A psychrometer is an instrument used for measuring the relative humidity of the atmosphere. It consists of two identical thermometers—one called a wet-bulb and the other a dry-bulb thermometer. In the former case the bulb is covered with a tight-fitting muslin cloth saturated with distilled water. The dry bulb thermometer acts as a control, measuring the actual temperature. The wet bulb thermometer will cool more rapidly because of the effect of the evaporating water from the cloth. The drier the air is, the greater the evaporation, which will result in decrease of temperature of the wet-bulb thermometer. Psychrometric tables should be consulted to determine various humidity variables. Sling psychrometers get their name because the thermometers are attached to the end of a sling (or chain). Whirling them by the chain will ensure proper ventilation and accurate readings.*

relative humidity. By measuring increases in humidity, meteorologists can predict impending storms.

The most common hygrometer is the dry- and wet-bulb psychrometer. It consists of two identical thermometers (electrical or mercury). One bulb has a moist cotton or linen wick wrapped around it. As the water in the cloth evaporates, the wick absorbs heat from the thermometer bulb, causing the wet thermometer reading to drop. In other words, the loss of water is cooling the air. With the help of a calculation table, you can use the difference between the reading from the dry thermometer and the reading from the wet thermometer to determine the relative humidity.

The most accurate way to measure humidity, however, is with an electric hygrometer. In electric hygrometers, a known volume of gas is passed over a hygroscopic, or moisture-absorbing, material such as phosphorus pentoxide. The difference in the weight of the gas before and after its passage over the moist material is used to determine how much water was taken out of the gas. Older types of hygrometer often use human hair (blond hair is preferred), which stretches as it absorbs moisture. Such older hygrometers are called mechanical because they rely on organic substances which contract or expand because of relative humidity.

Constructing a Mechanical Hygrometer

To construct a simple hygrometer to observe the principles at work, you will need:

1 A scrap piece of wood or flat Styrofoam about 9 inches (22 cm) long and 4 inches (10 cm) wide

2 A flat piece of plastic about 3 inches (8 cm) long and 3 inches (8 cm) wide, thin enough to cut through

3 Small nails

4 Long strands of human hair about 8 inches (20 cm) long

5 A dime or other small coin

6 Glue

7 Tape

8 Hammer

9 Scissors

Cut the piece of plastic into a triangle. Tape the dime onto the plastic, near the point. Hammer one of the nails through the plastic pointer, near the base of the triangle. Wiggle the nail until the pointer can move freely and loosely around the nail. Glue the hair strands to the plastic on the plastic pointer, between the dime and the nail hole.

Now position the pointer on the wood or Styrofoam base about three quarters of the way down the side. Attach the nail to the base. The pointer must be able to turn easily around the nail. Attach the other nail to the base about 1 inch (2 cm) from the top of the base, in line with the pointer. Pull the hair strands straight and tight so that the pointer points parallel to the ground, making the end of the pointer perpendicular to the hair. If you've done this correctly, the hair should be hanging perfectly vertical and the pointer should be pointing perfectly horizontal. Glue the ends of the hair to the nail, trimming it if necessary.

The hair will indicate the level of moisture in the air by expanding and contracting. Moist air will cause the hair to expand and lengthen, which will cause the pointer to point downward. When the air is dry, the hair will contract and shorten, causing the pointer to point upward.

Barometer

There are few more important weather markers than atmospheric pressure, so in the meteorological arsenal few instruments are more valuable than the barometer. By measuring increases or decreases in atmospheric pressure, a barometer can warn of impending storms or signal fair weather ahead. Falling barometric pressure indicates the approach of a low-pressure area associated with storms, while rising barometric pressure means a high-pressure system is moving in, bringing clearer weather.

The idea that decreasing pressure portends approaching storms has been known for centuries and was exploited by crude forecasting devices known variously as weather glasses, thunder glasses, or storm glasses, and sometimes as a "Goethe thermometer" because the great writer Johann Wolfgang von Goethe (1749–1832) popularized it in his native Germany. These early devices used water rather than today's mercury. The idea, though, has considerable validity: when air masses are displaced by a change in the course of jet streams or disturbances in the upper level of the troposphere (that is the layer of atmosphere where life flourishes and where weather occurs), winds rush in to fill the void, often lifting clouds, and producing condensation and precipitation—in other words, stormy weather. The pattern reverses itself with high-pressure systems.

The word "barometer" comes from the Greek words for "weight" (baros) and "measure" (metron). Although there are different types of barometers, they are all based on the same basic principle: measuring the weight of a vertical column of air through the atmosphere. Essentially, the barometer is a glass tube, from which the air has been removed, inserted into a dish of mercury. The mercury responds to the pressure of

Taking Readings

Make sure to check both surface and sea level readings: barometric pressure will vary depending on whether you are at the shore or at a higher elevation. Barometer readings during the year will average from around 29 inches to about 30.5 inches (74–77 cm). To forecast future weather conditions you should look for pressure changes—up or down—larger than one-tenth of an inch (0.25 cm). The faster the pressure change, the more severe the weather changes will be.

the air acting on it by moving up into the vacuum of the glass tube. At sea level, air pressure pushes the mercury about 30 inches (76 cm) up into the tube. Air pressure on the Earth's surface is usually given in inches of mercury; air pressure above is given in millibars, also known as hectopascals (hPa).

There are three basic types of barometer: mercury, aneroid, and digital:

A mercury barometer uses a vacuum tube of glass, which dips into mercury. Most mercury barometers are accurate to about one thousandth of an inch (0.0025 cm) of mercury. Although they are generally quite reliable, they are subject to pollution from the atmosphere and small errors can creep in that grow larger with time. Mercury is also a hazardous material.

An aneroid barometer measures the expansion of a metal cell or capsule that contains a partial vacuum. The cell will contract under high pressure, causing a pointer attached to it to bend forward to give a high-pressure reading on the face of the instrument. Because it relies on a mechanical assembly, this type of barometer is more susceptible to error and, unless it is checked about once a month, it will become more and more error-prone.

A digital barometer is the most accurate and also the most expensive. It uses electronic circuitry to detect changes in pressure and then to indicate them digitally. Its accuracy is consistently within a hundredth of an inch (0.025 cm).

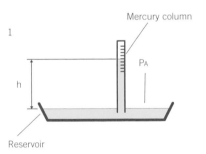

1

Mercury column

P_A

h

Reservoir

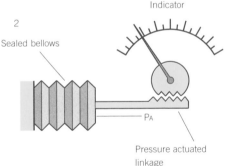

Indicator

2

Sealed bellows

P_A

Pressure actuated linkage

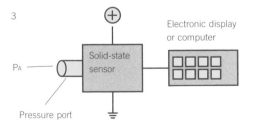

3

Solid-state sensor

P_A

Pressure port

Electronic display or computer

P_A = atmospheric pressure

Above: *Earlier models like the mercury (1) and bellows (2) barometers were limited because of errors that crept into the readings over time. Today digital (also solid-state) (3) barometers provide the most accurate information about changes in air pressure.*

Anemometer and Wind Vane

An anemometer is designed to measure wind velocity and pressure, while a wind vane or weather vane measures the direction of the wind—which always refers to the direction from which it is coming, not the direction in which it is blowing.

Anemometers

Anemometers measure wind velocity and pressure, but because wind velocity and pressure are related, information about one will also provide information about the other. The word "anemometer" comes from the Greek word "anemos," which means wind. Invented by Leone Battista Alberti (1404–72), anemometers come in a variety of types. The simplest type has three or four hemispherical cups mounted on each end of a pair of horizontal arms, which lie at equal angles to each other. The cups are designed to catch the wind and cause the instrument to rotate around a vertical axis.

Below: *The rotating blades of this computerized anemometer indicate wind speed. The data is fed into a computer, allowing precise real-time measurements.*

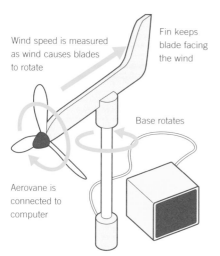

Wind speed is measured as wind causes blades to rotate

Fin keeps blade facing the wind

Base rotates

Aerovane is connected to computer

By calculating the number of turns made at any given time, it is possible to determine the velocity of the wind.

Sonic anemometers rely on sound waves to measure wind speed and direction; because of their precision and lack of moving parts they are used by many automated weather stations. Pressure anemometers—which, as the name suggests, measure wind pressure—are categorized as either a plate or a tube anemometer. In the plate type, a flat plate, kept exposed to the wind by a wind vane, is balanced by a spring. Wind pressure is calculated by the measure of the spring's distortion, which is registered on a gauge. Plate anemometers are limited in that they do not perform well in recording high or variable winds. Tube anemometers are in the shape of a U tube containing liquid at one end; depending on whether the wind blows into the mouth of the tube or horizontally over it, the pressure in the liquid will respectively increase and decrease. Because these differences are very slight, it requires special means to record them. Tube anemometers have the advantage that they can be mounted on a high pole and left alone for years without any need for maintenance.

Wind Vanes

Wind vanes, also known as aerovanes and weather vanes, are movable devices attached to an elevated object—often a roof—intended to show the direction of the wind. Since they are frequently fashioned in the shape of cockerels, they are often called weathercocks. Wind vanes need to be balanced on either side of their axis, but the parts that are exposed to the wind must be unequal. It is that inequalilty that makes the device rotate: the smaller area of the vane is set into motion by the need to minimize the impact of the wind on its surface. That's the part that has an arrow pointing toward the source of the wind. Most simple weather vanes indicate the direction of the wind by markers beneath the pointer.

Building Your Own Wind Vane
To build your own wind vane, you will need the following materials:

1 Sturdy ruler or piece of wood 12 inches (30 cm) long
2 Aluminum pie plate

Aluminum baking dish

Tail

Nail

Washer

Shaft of broom handle

Head

3 A long wooden dowel (about the size of a broom stick)
4 Nails
5 Metal washer
6 Hammer
7 Glue
8 Small saw (or serrated knife)
9 Wire (for mounting)
10 Scissors (strong enough to cut aluminum)

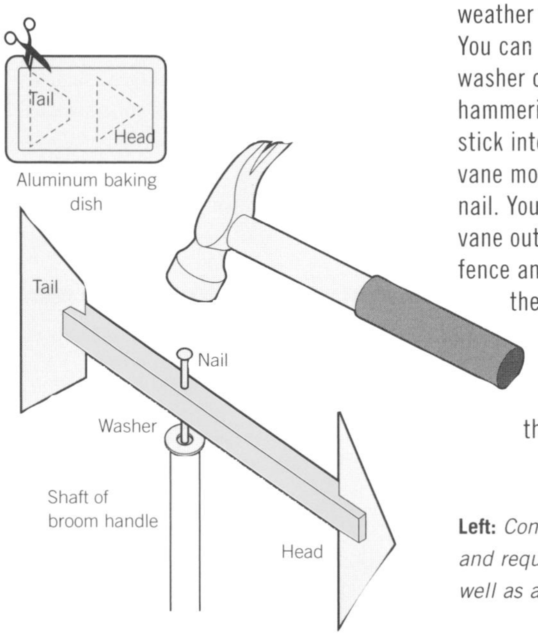

With a saw or a pair of sharp scissors, make a vertical slit at each end of the ruler about 1/2 inch (1.5 cm) deep. Hammer one nail all the way through the ruler at its midpoint (exactly halfway). Keep turning the wood around the nail several times until the ruler turns easily. Now cut the head and tail for your wind vane from the aluminum plate, as shown in the diagram. Glue the head into the slot at one end of the wooden stick and the tail into the other end.

Once the glue is dry, attach the weather vane to the long wooden dowel. You can do this by placing the metal washer on the end of the dowel and then hammering the nail through the wooden stick into the dowel. Make sure that the vane moves freely and easily around the nail. You can now mount the weather vane outside. You may want to use a fence and secure it with wire. Try to get the vane as high above the fence as possible while making sure it is still well secured. The head of the pointer will always indicate the source of the wind.

Left: *Constructing a weather vane is simple and requires only a few simple items as well as a good hand to wield a hammer.*

Rain Gauge

Rain gauges, which collect rainfall, are commonly used at weather stations and airports to measure accumulation. Care must be taken to ensure that the gauge is protected from splashing, which could distort the findings. Measuring snow is trickier because some of it may melt before an accurate determination can be made.

The Koreans can probably take credit for the world's first rain gauge, developed in 1441 under orders of King Se Jong. He had rain guages sent to every village to assess each farm's potential harvest—the estimates then helped the government figure out what each farmer's land taxes should be. The standard rain gauge in use today by meteorologists and airports, however, was only invented a century ago. It consists of a cylinder 20 inches (50 cm) high and 8 inches (20 cm) in diameter, with a funnel and inside it a smaller measuring tube. The funnel directs the precipitation into the measuring tube, which is marked in increments of tenths and hundredths of an inch (0.25 and 0.025 cm), allowing meteorologists to make precise measurements. When water reaches the 30 marker, for instance, you would say that you had 3 tenths (3 x 0.25 cm) or 30 one-hundredths (30 x 0.025 cm) of an inch of rain. When rain is less than 1 hundredth of an inch (0.025 cm), meteorologists refer to the rainfall as a "trace" of rain.

The standard rain gauge can measure up to about 2 inches (5 cm). In the event of a particularly heavy rainfall, water will overflow the measuring tube, spilling into the cylinder containing it, which is known as the overflow tube. The problem is resolved simply: the observer empties the measuring tube and then pours the residual water into it; all that remains is to add the two totals. Another problem can develop when the temperature is close to or below freezing, since ice or snow may form, impeding the collection of any subsequent rain. Rain gauge amounts are read either manually or by sensors at an automated weather station. In some countries, professional meteorologists are supplemented by a network of volunteers who collect precipitation data in sparsely populated areas.

Taking Measurements and Positioning

Although reading a rain gauge is fairly simple, mistakes can occur. As rain fills the measuring tube, the surface of the water is distorted: it becomes curved because of the surface tension of a liquid. The curve is called a meniscus. You should read the base of the meniscus for a true reading.

Problems in accuracy can result from the position of the gauge. Even the simplest type of rain gauge will perform better out in the open than a more sophisticated gauge placed under a tree. The rule of thumb says that you should locate the gauge at a minimum distance of twice the height of nearby obstructions, meaning that the gauge should be kept 30 feet (10 m) from a structure 15 feet (5 m) tall.

A grassy area is better than a hard surface. A hard surface will be prone to splashing, which can cause false readings. Gravel or pebbles should be placed in a circle around the gauge to prevent any damage if someone is mowing the grass.

Wind screens—large boards that protect rain gauges from winds—may also be warranted in regions where snow comprises about 20 percent of annual precipitation. They should be placed at least 1 foot (30 cm) from the gauge to prevent eddy currents from forming that could interfere with rain collection. Winds will cause more losses in precipitation catch during a snowfall than a rainfall.

Choosing Rain Gauges and Snow Boards

When you look for a rain gauge there is one basic rule of thumb: the larger the gauge, the more accurate the catch. That's because a larger gauge is capable of taking a more representative sampling of the precipitation. Larger gauges have an additional advantage in that they are better at collecting drizzle and snow, although snowfall measurement can be distorted by eddies forming around the edges of the gauge. The largest and most accurate rain gauge is the 8-inch (20-cm) gauge. It has a large overflow can and measuring tube capped loosely by a funnel. Excess water from underneath the funnel can leak into the overflow. To measure the rainfall you should use an absorbent, calibrated stick which is dipped into the measuring tube. The 4-inch (10-cm) gauge is not as accurate. Although it is constructed in the same way, it is usually transparent, allowing you to read the amount of rainfall against marks on its side. Tube gauges, which are around the size of a test tube, are so unreliable that most experts recommend not to use them.

Wedge gauges, which are quite small and slender, are readily available in department stores, but their accuracy is problematic. To read the meniscus takes good eyesight and patience. Wedge gauges are useless to collect snow, so you should use a snowboard instead. This is basically a flat platform for snow to fall upon. You merely lift the snowboard and brush the snow off to determine the amount of new snowfall. A snowboard should be made of wood or Styrofoam and securely anchored on the ground. It should be painted white to minimize melting.

Rainfall Facts

Rainfall is unevenly distributed around the Earth, influencing climate, agriculture, and animal life. In deserts annual rainfall is usually 1 inch (2.5 cm) or less, while in the Hawaiian mountains and the hills of Assam in northeastern India it can be as much as 400 inches (1,000 cm). In the United States, rainfall ranges from less than 2 inches (5 cm) in Death Valley, California, to in excess of 100 inches (250 cm) on the coast of Washington State. In the U.K. annual rainfall exceeds 60 inches (150 cm) and in places can reach 200 inches (500 cm). In Australia, rainfall is highest in the northern coastal regions, where it reaches 125 inches (320 cm) and lowest in the interior, where only trace amounts occur.

The Future of Amateur Weather Forecasting

The Internet has transformed what was once a hobby into a formidable data-sharing network maintained and monitored by thousands of amateur weather enthusiasts around the world, resulting in more accurate forecasts for everyone.

Personal weather stations are already performing a valuable service by recording data in locations that are not monitored by professional services or agencies. In parts of the United States, for instance, official weather stations may be over 40 miles (70 km) from one another. By contrast, private weather stations are often within a few miles of one another. Accurate forecasting depends on two basic factors: geography and time. Weather forecasting based on conditions monitored over a large scale—a region, for example—will not be as accurate as a forecast based on conditions observed at several points within that region. By the same token, if the information is not produced and analyzed in a timely fashion it will be of little practical use.

Computer weather forecasting models, employed by meteorological agencies, are based on a division of the Earth into squares or grids. As computer models have advanced, the squares have diminished in size—in other words, meteorologists are able to analyze conditions at smaller and smaller scales. They are also trying to update their data more frequently. Weather conditions can vary greatly even within short distances and they can change rapidly: for example, a thunderstorm can appear and vanish within three hours. Amateurs are filling in the gaps in both respects. "We've always struggled to get information from out there, and it's been great," noted James Scarlett, the warning coordination meteorologist for the National Weather Service in Montana, United States. "The more data we get from the field, the better our forecasting."

Sharing Data

Professional meteorologists are increasingly relying on contributions from amateurs to provide a bigger picture of weather patterns than was previously possible. Amateurs are also forming online associations that gather data and then forward them to national agencies. In the United States, for instance, the Citizen Weather Observer Program (www.cwop.net) feeds data to the National Oceanic and Atmospheric Administration, the umbrella agency for the National Meteorological Service. The group's data are routinely used by 80 percent of its forecasting offices.

Also in the United States, the National Weather Service (NWS) Cooperative Observer Program (www.nws.noaa.gov/om/coop) is a network of more than 11,000 volunteers who collect weather data, such as daily maximum and minimum temperatures, snowfall, and 24-hour precipitation totals. Equipment to gather these data is provided and maintained by the NWS. Observers send data forms monthly to the National Climatic Data Center, where data are digitized, checked, and archived.

A loosely knit group called Weather Underground (www.wunderground.com) claims a worldwide membership of over 12,000 personal weather stations and receives over 3,200 uploads to the site daily. Amateurs are welcome to join the network.

Schools are getting in on the act as well. WeatherBug, a software program supported by advertising, relies on a network of 7,000 weather stations maintained by schools in the United States; the program sits on millions of desktop computers throughout the country and its information is used by dozens of TV stations.

These amateur weather station networks can pinpoint phenomena such as tornadoes, and intermittent rain and snow lines in winter storms that often escape the attention of official meteorological surveillance. In winter, for instance, some localities may be getting snow while a few blocks away people are experiencing rain. Without the input from amateurs who actually live in these localities, the meteorological agencies and TV and radio stations would have no way of knowing. An authoritative announcement that snow is coming down heavily is not likely to add to a forecaster's credibility when his listeners only have to look out the window to see that it's raining.

Electronic Weather Stations

Several integrated weather stations are available at a range of prices. The following is a general guide to possible packages:

• A variety of weather stations can be obtained from Davis Instruments, such as the Vantage Pro2. It includes a full range of sensors with a wireless connection to an indoor console reader.

• Oregon Scientific makes wireless weather stations such as the WMR968. It includes a thermo-hygrometer, rain gauge, and anemometer.

• For an inexpensive basic weather station, Peet Bros. offer products such as Ultimeter 100. It can be linked to the most common weather sensors and is supplied with an adjustable flash-flood alert. The station also has the capacity to transmit weather data by phone, modem, or radio link.

The following are software packages that can be used alongside compatible weather stations. They allow for the transmission of data from instrument to computer, use graphs to chart weather trends, and offer interface graphics. They also have the capacity to upload information to a Web site or a community weather service.

• Virtual Weather Station from Ambient Weather is a popular weather program that converts data into on-screen bar graphs and meters. An Internet Edition allows information to be uplinked to Web sites.

• Weather View 32 from weatherview32. com allows you to add Internet displays such as satellite images to personal weather station data. It can send weather alerts to e-mail addresses and pagers.

Predicting Weather from Clouds

To obtain clues about changes in the weather, you can study the clouds to see what form they take, what height they form at, whether they are opaque or transparent, and whether they are moving. There are five different classifications of clouds that are based on their characteristic properties, especially altitude: high-level, mid-level, low-level, convective (or vertically developed), and all other clouds.

Classifying Clouds

Cloud patterns constitute a kind of language; once you are able to interpret this language you will be in a better position to determine what kind of weather to anticipate in the near future.

Clouds are classified by their size, formation, and by the height of their cloud base. Their names are derived from the Latin words used to describe their appearance from the ground. "Cumulus," for instance, comes from the

1 Warm ground

2 Hills and mountains

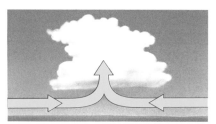

3 Air from opposite directions

4 Cold air mass

Above: *Clouds form from rising air when: (1) the Sun warms the ground and creates thermals; (2) air moves up over a hill or mountain; (3) air from opposite directions meets and is forced upward; (4) cold air flows under and lifts warmer, lighter air.*

Right: *Cloud names relate to their height and shape. The three major classifications are high-level, mid-level, and low-level, although some clouds, such as cumulonimbus, may extend from the bottom to the top of the troposphere.*

word for "heap," which is an accurate description of the puffy cotton-candy appearance they assume. "Stratus" means layer, "cirrus" means curl of hair, "nimbus" means rain. There are also subcategories based on the altitude of the cloud base, for example clouds with the prefix "alto" are usually formed at middle levels.

Cloud Watching

Clouds are a vital part of the warming and cooling system of the planet. Clouds help keep the Earth warm, acting as an insulating blanket. On the other hand, clouds block much of the Sun's energy and radiate heat back into the atmosphere. Water vapor in clouds radiates more heat than gases in clear air. How much energy-radiating and energy-absorbing matter will be found in a particular cloud depends on its shape and size. Stratus clouds, which are composed of smooth sheets of water vapor, tend to reflect more sunlight than they absorb, producing a cooling effect on the ground below that in turn results in condensation of water vapor and rain. Cirrus clouds, composed of ice crystals, permit more sunlight to penetrate and are associated with a warming effect.

When making your observations, do not limit your cloud-watching to just one mass of clouds in a particular part of the sky: take into account the variety of different cloud formations, because any change in atmospheric conditions in one place is likely to have an effect on conditions in another. You should also take note of how much of the sky is covered by clouds. Meteorologists measure cloud cover in units called oktas, with each okta representing one eighth of the sky covered by cloud, ranging from clear skies (0 oktas) to completely overcast (8 oktas).

High-Level Clouds

High-level clouds form above 20,000 feet (6,000 m), an altitude where temperatures are very cold. As a result, the water vapor freezes, forming clouds mainly composed of ice crystals. The names of these clouds are distinguished by the prefix "cirr." They tend to be thin and white in appearance, although they may light up with color at sunset.

Cirrus

Cirrus clouds are associated with the approach of cold or warm fronts and can signal the approach of stormy weather. In the middle latitudes, cirrus clouds generally move from west to east, propelled across the sky by strong westerly winds. Cirrus clouds take on different forms depending upon the strength of the wind. Cirrus fibratus is the most common type, and looks like fibers or feathers. Cirrus uncinus, also known as mares' tails, exists as long, sweeping lines of cloud that have a hook at the end, like the tip of a ski. Pileus clouds are layered cirrus clouds lying on top of other clouds, or over mountaintops.

Cirrocumulus

Cirrocumulus clouds are small, rounded white puffs that are isolated or in long rows. When cirrocumulus clouds form rows, it is their characteristic rippling appearance that distinguishes them from cirrus or cirrostratus clouds, and gives them their common name, "mackerel sky," because they resemble fish scales. These clouds rarely cover an entire sky. They are among the most spectacular clouds in the sky because of their ability to capture the red and yellow light of the setting sun.

Cirrocumulus clouds form from cirrus clouds that are warmed from below. The heating sets in motion convective currents—pockets of air that rise and sink inside the cloud. Cirrocumulus clouds usually appear in the winter and indicate fair but cold weather. In tropical regions, however, they may indicate an approaching hurricane.

Cirrostratus

Cirrostratus clouds are sheetlike in appearance. These clouds, which can build up to a thickness of several thousand feet, can blanket an entire sky. In spite of their thickness, they are still relatively transparent, allowing an observer to view the Sun or Moon through them. Cirrostratus clouds may also form haloes around the Sun or Moon, resulting from the refraction

Above: *Cirrus uncinus; pileus capping a cumulus congestus cloud; cirrus fibratus.*

Above: *Cirrocumulus.*

Above: *A haloed cirrostratus.*

of light by the cloud's ice crystals. Refraction occurs when light passes through mediums of different densities.

These high-level clouds typically form from a process known as convergence. Convergence occurs when there is a large horizontal inflow of air into a region on the Earth's surface. That air needs somewhere to go and, since it cannot go downward, it rises. This lifting process produces cirrostratus clouds. Their appearance can indicate a warm front and approaching precipitation in high latitudes; the thicker the clouds, the greater the chance of precipitation within 15 to 25 hours or sooner. In tropical and subtropical latitudes, where no warm fronts occur, the presence of cirrostratus clouds means that there is moisture in the upper troposphere but does not signal rain.

Cirrus Aviaticus

Cirrus aviaticus are wispy streaks of cloud that form at high altitudes, approximately 33,000–39,000 feet (10,000–12,000 m), due to the temperature and pressure changes produced by jet engines. The clouds' name, from the Latin, means "aviation cloud."

Ordinarily, temperatures in this region of the atmosphere are extremely cold (-40°F/-40°C). Jet aircraft leave behind a heated mixture of water and particulates (byproducts of the combustion of fuel) made up of carbon dioxide, nitrogen oxides, sulfates, and soot. These particulates form a nucleus for warm water droplets to condense around before quickly freezing into ice crystals. These contrails, as they are sometimes called, can disappear within seconds or last several hours depending on prevailing atmospheric conditions. Strong winds will quickly spread the streak. If they remain visible, the air is relatively humid.

Above: *Cirrus aviaticus, formed by jets.*

Mid-Level Clouds

These clouds, identified by the prefix "alto," are composed of both water droplets and ice crystals. They are typically thin and are distinguished from higher clouds by their shaded gray or blue-gray appearance. Mid-level clouds are found between 6,500 and 23,000 feet (2,000–7,000 m).

Altocumulus

Altocumulus clouds are puffy patches that sometimes form wavelike masses or parallel bands. A portion of these clouds is typically gray-shaded, a characteristic that sets them apart from the high-level cirrocumulus. They are usually thin, less than 300 feet (100 m) thick. If you hold your arm up to the sky they are about the size of your thumbnail. They tend to blow whichever way the wind is blowing—or, more technically, they may be oriented along any wind-shear direction at that altitude.

Altocumulus clouds may signal the advance of a cold front, as the cooler air lifts the warmer air, producing condensation. The appearance of altocumulus clouds generally suggests that rain or overcast skies will follow within 15 to 25 hours. On a hot, humid morning in summer, if you spot altocumulus clouds in the sky, thunderstorms may result later in the day.

Altostratus

Altostratus clouds, grayish or blue-gray in appearance, form a layer that can cover the entire sky. They can extend up to hundreds of miles horizontally and up to thousands of feet vertically. These clouds—striated, fibrous, or uniform sheets—are generally featureless. Occasionally they appear as undulations or in parallel bands. They are made up of two or more superimposed layers of supercooled and frozen particles of water. (Supercooled means that the water drops don't freeze: they just became very cold.)

Above: *Altocumulus.*

Because of their density, altostratus can be distinguished from cirrostratus clouds by the capacity of even the thinnest layers to partially or completely obscure the Sun. Consequently, they never produce the kind of halo effect around the Sun or Moon that altocumulus clouds often do.

Altostratus frequently appear in advance of storms and typically produce snow crystals and drizzle. Often the precipitation evaporates before reaching the ground. Altostratus clouds form when a front of warm, moist air meets a body of cold, dry air. The collision of these two air masses often sets the stage for precipitation. The thicker the clouds, the more likely are the chances of rain and snow.

Altostratus clouds may be accompanied by clouds in ragged shreds that form at a considerable distance beneath their undersurface in turbulent layers. These clouds are called plannus

Above: *A layer of altostratus clouds typically covers the entire sky.*

clouds and make their debut in the sky as small, sparse, and well separated. Plannus clouds are formed from the evaporation of precipitation from the altostratus clouds. As the altostratus clouds thicken and their base begins to lower, the distance separating them from their daughter plannus clouds diminishes, causing the ragged plannus clouds to proliferate and grow in size until the two types of clouds sometimes merge into one continuous layer.

Cloud Classification

The current classification system for clouds dates back over 200 years. It was developed in 1803 by Luke Howard, an English meteorologist, who divided clouds into three categories—cirrus, stratus, and cumulus—based on observations from the ground. That system has since been enlarged to include several different subtypes based on various properties, including shape, altitude, and internal structure. But now some scientists are saying that the

system requires modification. That is because, since the 1960s, meteorologists have increasingly come to rely on satellite imagery rather than ground observations to observe clouds. Moreover, satellites are providing data on microwave or infrared emissions from clouds that were previously unavailable. For instance, a wispy cirrus cloud over a snow-covered ground is difficult to detect in visible light but shows up clearly in infrared imagery.

Low-Level Clouds

Low-level clouds are found below 6,500 feet (2,000 m). They are generally composed of water droplets, but when temperatures are cold enough, they may also contain ice particles and snow. The appearance of these clouds often portends precipitation, although they seldom produce heavy downpours.

Nimbostratus

Nimbostratus clouds are dark and gray, covering the sky and blotting out the Sun. They are almost invariably associated with continuous light to moderate precipitation over large areas. The prefix "nimbo" means rain. Threatening as they may appear, nimbostratus almost never produce heavy downpours. However, they can cause rain lasting several hours to more than a day. They can extend high enough into the atmosphere for ice crystals to coalesce into snow, but the snow will generally melt before it reaches the ground.

These clouds tend to form in a stable atmosphere where, over a relatively large area, warm, moist air is overrunning cooler air at the surface. This results in the lifting and condensation of the warm air. The clouds that form below nimbostratus clouds, as fog and fast-moving "scud" or fractostratus clouds, reduce visibility.

Stratus

Stratus clouds are uniformly gray clouds that extend from horizon to horizon. They are paler than nimbostratus. Stratus clouds can form when very weak updrafts lift a thin layer of warm air high enough to cause condensation. They can also form when a layer of air is cooled from below to its dew point temperature and the water vapor condenses into liquid droplets. Stratus clouds seldom produce precipitation because the updrafts that produce them are so weak.

Stratocumulus

Stratocumulus clouds are low, lumpy, puffy clouds that occur in rows, patches, or rounded masses, often with patches of blue sky between them. Stratocumulus range from white to dark gray. If you hold your arm up to the sky they seem to be about the size of your fist. These clouds can occur simultaneously at two or more

Above: *Nimbostratus clouds blot out the Sun and bring rain.*

Above: *Stratocumulus clouds appear in rows or patches and are rarely associated with rain.*

levels, and their size, thickness, and shape can vary considerably. They may be sufficiently transparent to discern the Sun through them—they can sometimes form a corona around it—or they can be entirely opaque. They are no more than half a mile to a mile (1–2 km) thick, but on very rare occasions, if the cloud is frozen, it can form into what is called a sun pillar.

These clouds are rarely associated with precipitation, although occasionally they bring precipitation—rain, snow, or ice pellets—of weak intensity.

Fractocumulus

Fractocumulus look like disorganized puffs of cumulus clouds that pilots sometimes refer to as "scud," although that term is more often applied to fractostratus. They are often remnants of other clouds, particularly cumulus, that have dissipated, or else form in the outflow beneath a thunderstorm.

Above: *Fractocumulus clouds have a characteristically disorganized, puffy appearance.*

Convective Clouds

Convective clouds form in association with convection, the rapid rising of parcels of air, often formed when the Sun warms the earth through the day. They assume a variety of forms—mounds, domes, or towers— but always develop vertically. Towering convective clouds, known as cumulonimbus, can reach to 11 miles (17 km) high, nearly to the top of the troposphere, and are associated with severely unstable weather.

Convective clouds are brilliant white on top, where they are exposed to sunlight, and dark below. Because of their vertical development, they may grow in a range of temperatures, from above to below freezing, and can be composed of water droplets, supercooled water droplets, ice crystals, snowflakes, or a combination. All convective clouds start life as a cumulus humilis.

Cumulus Humilis

Cumulus humilis are small clouds with minimal vertical development and may have a flattened appearance. "Humilis" means "humble" in Latin. They look like sharply defined, white pillows. Cumulus humilis never produce precipitation and are sometimes called fair-weather cumulus. They appear in an otherwise clear sky on warm days, in late morning or early afternoon, and

are caused by the rising of air heated by the Sun. They are formed in updrafts of air and surrounded by downdrafts. Individual clouds commonly come and go in less than an hour—although, under the right conditions, they may develop into cumulus mediocris. As the heat of the day dies, so will cumulus humilis.

Cumulus Mediocris

Cumulus mediocris clouds are formed when convection is stronger than in the case of cumulus humilis. They often form in spring, when the Sun warms up the land below cold air, making the air unstable. Cumulus mediocris show some moderate vertical development, with small protuberances and sprouts at their tops. They typically produce no precipitation. Mediocris are the transition clouds between cumulus humilis and cumulus congestus.

Cumulus Congestus

Cumulus congestus clouds develop from cumulus mediocris clouds, when strong currents of hot air are rising high into the atmosphere, creating unstable conditions. These clouds resemble a cauliflower, with more prominent sprouts than a cumulus mediocris, and are sharply delineated. These clouds can develop vertically very quickly and can produce considerable precipitation, typically rain showers.

Above: *Cumulus humilis clouds show very little vertical development.*

Above: *Cauliflower-shaped cumulus congestus typically produce showers.*

Cumulus Castellanus

Cumulus castellanus are cumulus congestus clouds that have taken on the form of a narrow, very high tower, which is why they are often called towering cumulus. The tops of the towers may be formed of small puffs that can separate from the cloud; carried away by the wind they can sometimes form virga—small undulating clouds—or dissipate completely.

Cumulonimbus

Cumulonimbus clouds result from the development of cumulus congestus clouds. Within 15 minutes, the cauliflower tops of a congestus cloud can develop into the anvil-shape of a cumulonimbus. These heavy, dense clouds assume the form of mountains or huge towers. Their upper portion may be smooth, fibrous, or striated, but is almost always flattened into the shape of an anvil or plume. The anvil-shape is caused by strong upper-level horizontal winds and by updrafts of warm air. The base of a cumulonimbus cloud can be very dark. They are often accompanied by low, ragged clouds. Cumulonimbus clouds are known as cloud factories because they can produce other types of clouds—altocumulus, altostratus, or stratocumulus—as parts of their upper portion branch out on their own.

The formation of cumulonimbus clouds is abetted by topology: when they form near oceans they draw energy from moist sea air; when they form over mountains they are fed by air pushed upward by the peaks. Cumulonimbus are likely to form on summer afternoons as the Earth's surface heats up. In addition, they can also form along cold fronts when warm air is forced up by cold air.

These clouds are associated with violent thunderstorms and tornadoes. Torrential rain or hail showers are characteristic. If you can judge solely by observing the undersurface, cumulonimbus may be mistaken for nimbostratus clouds when they cover a large expanse of sky. In that case, the best way to distinguish is by the intensity of the precipitation generated. Steady, continuous rain is more likely to come from nimbostratus clouds.

Above: *An anvil-shaped cumulonimbus.*

Other Types of Clouds I

Some types of clouds defy easy classification but are fascinating for the weather watcher. Mammatus clouds, which are linked with cumulonimbus clouds, look like pouches, while roll clouds have a horizontal tube shape. Each type is associated with a distinct weather pattern, so they are well worth familiarizing yourself with.

Mammatus

Mammatus clouds are deceptive: forming on the underside of cumulonimbus clouds, they appear more threatening than they actually are. Paradoxically, they indicate that a storm system is weakening. They resemble pouches with breastlike appendages and tend to protrude from the bottom of a thunderstorm's anvil cloud. (Their name is derived from the Latin meaning "breastlike.") They typically occur on the rear side of the storm although they can be seen at the leading edge as well.

Mammatus clouds only occur in the presence of thunderstorm-producing cumulonimbus clouds but they can drift several miles away from the storm itself. In order to form, these clouds need moist, unstable conditions in the middle to upper atmosphere over a dry lower layer. These conditions cause updrafts to occur, which sculpt the cloud into its characteristic pouchlike shape. Meteorologists now believe that these clouds are probably byproducts of tornadoes, not a precursor of them.

Roll Clouds

Roll clouds, technically known as arcus clouds, are often associated with storms. They usually have a low, horizontal tube shape and take their name from the way in which they seem to roll about a

Above: *Mammatus clouds typically occur as a storm is weakening.*

Above: *Roll clouds are pushed ahead of a storm by strong gusts.*

horizontal axis but not at a rapid pace. They are produced when warm air is lifted up in front of a storm system and are propelled by strong gusts from the outflows of air generated by the thunderstorm. They are relatively rare and have a menacing appearance. Roll clouds are completely detached from their parent cloud, which distinguishes them from shelf clouds (see p. 82).

Cloud Types and Weather

Cloud Type	Description	Weather
High-level Clouds		
Cirrus	Thin and wispy, they can appear in a variety of shapes.	Associated with an approaching front and stormy weather.
Cirrocumulus	Small, rounded white puffs that are isolated or positioned in long rows.	These winter clouds indicate fair but cold weather.
Cirrostratus	Sheetlike clouds covering the entire sky.	May signal precipitation within 15–25 hours.
Cirrus aviaticus	Wispy streaks.	If they linger, the air is relatively humid.
Mid-level Clouds		
Altocumulus	Puffy clouds in bands or wavelike masses.	May signal precipitation within 15–25 hours.
Altostratus	A thin layer of grayish or blue-gray clouds that can cover the entire sky.	Frequently appear in advance of storms and typically produce snow crystals and drizzle.
Low-level Clouds		
Nimbostratus	Dark, low, uniformly gray clouds.	Widespread light to moderate precipitation.
Stratocumulus	Low, lumpy, puffy clouds occurring in patches or in rounded masses.	Sometimes accompanied by precipitation—rain or snow—usually of weak intensity.
Fractocumulus	Unstructured and disorganized puffs of cumulus clouds.	Produced as a result of a thunderstorm.
Convective Clouds		
Cumulus humilis	Small puffy clouds with minimal vertical development.	Associated with fair weather and never produce precipitation.
Cumulus mediocris	Puffy clouds with moderate vertical development and small protuberances and sprouts at their tops.	Associated with fair weather and typically produce no precipitation.
Cumulus congestus	Resemble a cauliflower with prominent sprouts and a sharply delineated outline.	Produce considerable precipitation, typically showers.
Cumulus castellanus	Resemble narrow, very high towers with the tops formed of small puffs.	Suggest the onset of stormy weather.
Cumulonimbus	Heavy, dense clouds in the form of mountains or huge towers, often with an anvil-shaped top.	Produce heavy precipitation in the form of showers with thunder and lightning or hail.
Other Clouds		
Mammatus	Pouches that protrude from the bottom of a cumulonimbus cloud.	Indicates that a storm system is weakening.
Roll cloud	Low, horizontal tubelike shape and seem to roll about a horizontal axis.	Associated with storm systems.
Shelf cloud	Usually curved or semicircular and protruding from a cumulonimbus.	Indicates an advancing cold front.
Funnel cloud	Rotating column of air that extends from the base of a cloud (typically cumulonimbus or cumulus castellanus).	Very weak tornadoes with the potential to turn into full-blown tornadoes, often associated with supercell thunderstorms.
Wall cloud	Ragged, dark with signs of weak rotation.	Develop after an intense thunderstorm.

Other Types of Clouds II

Other cloud types worth noting include shelf clouds, which are associated with thunderstorms, and funnel and wall clouds. A cloud that forms close to the ground is known as fog. Fog and mist both consist of water vapor. The difference between fog and mist is simply one of density: because it contains more water droplets, fog is denser than mist.

Shelf Clouds

Shelf clouds, associated with cumulonimbus clouds, are low and horizontal. Usually curved or semicircular, they protrude from their parent cloud just like a shelf. Their appearance typically indicates an advancing cold front, where the warm air is being forced out and upward. They can be smooth or ragged depending on the stability of the air. Shelf clouds are also known as "front-loaders" because they are pushed in front of the storm as it gains strength.

Funnel Clouds

Funnel clouds are very weak tornadoes that do not touch the ground. For a tornado to occur, a funnel cloud must form first. Funnel clouds are defined as any rotating column of air that extends from the base of a cloud (typically cumulonimbus or cumulus castellanus) without reaching the ground. They can be cone-shaped in appearance or resemble a needlelike protuberance emerging from the main cloud base.

Wall Clouds

Wall clouds are ragged, dark, and show signs of weak rotation. Their origin is not completely understood but it is believed that they develop after an intense thunderstorm when air cooled by the rain is pulled upward, dragging more buoyant air along with it.

Above: *Low, horizontal shelf clouds are associated with thunderstorm activity.*

Fog and Mist

Pilots define fog and mist by visibility: if they cannot see farther in mist than 3,000 feet (1,000 m), it is fog; for the earthbound motorist it becomes fog if visibility is below 650 feet (200 m). Fog can form in a number of different ways, but it will usually form when moist air close to the ground is cooled, causing the water vapor to condense into droplets.

The temperature at which condensation occurs is called the dew point. If the air is cooler, it is able to hold less moisture and condensation occurs. When air is heated by the Sun it acquires a greater capacity to hold moisture, and the Sun "burns off" the fog. This is why fog usually dissipates shortly after dawn. That process is reversed at night when the ground loses heat. Temperatures tend to drop lower on long, winter nights when the sky is clear, creating conditions for dense fog to form.

Types of Fog

Precipitation fog occurs when rain or snow falls. The precipitation falls into drier air below the cloud and evaporation occurs, producing water vapor. This water vapor increases the moisture content of the air while cooling it. As the air below the cloud becomes saturated, fog will begin to form.

Advection fog forms when warm, moist air travels over a much colder area of land or sea. The air is cooled and, if the air is close to saturation, the moisture will condense and form fog. This phenomenon is usually observed in coastal regions, especially in spring and summer when temperature difference between sea and land is typically larger. Advection fog also occurs inland during thaws in winter or early spring. Advection is defined as the horizontal transport of any meteorological element, such as humidity or temperature. It is this kind of fog that is often blamed for car pileups and delays at airports.

Radiation fog often forms in autumn as nights become longer. These longer nights give the ground more time to cool, which in turn hastens the process of condensation of water vapor. It is called radiation fog because the ground radiates heat into the air. It is also known as ground fog because of the fog's tendency to hug the ground.

Steam fog can occur after a rain shower. It forms when the ground is sufficiently warm to promote evaporation. If, however, the air above the surface is saturated and is incapable of holding any more water, the excess moisture condenses in a manner that makes it look like steam. This effect can be seen at the poles, when cold air sweeps over a slightly warmer sea.

Upslope fog occurs on large hills and on mountains. This type of fog forms when winds blow up the slopes and cool the air.

Valley fog, as its name implies, is found in mountain valleys, particularly in wintertime. River valleys are particularly congenial to fog formation, because the river contributes additional moisture. A valley fog can be very dense and can persist for days at a time because the Sun is too weak to warm the air during the day sufficiently to dispel the fog. In this way the fog becomes even thicker when night falls and the air cools once more. Moreover, cool, dense air tends to pool at the bottom of valleys. The fog will linger until wind strong enough to blow it away barrels through the valley.

Predicting Precipitation

Threatening clouds may portend rain and send people scurrying for cover, but how do meteorologists forecast rain on days when the skies are clear? What form will precipitation take on days when temperatures are just below or just above freezing? Complicating matters, forecasters must also take into consideration conditions in the atmosphere before they can predict what will come down from the sky and where.

When cloud particles—water droplets, supercooled water droplets, or ice pellets—become too heavy to remain suspended in the air, they fall to the ground as precipitation. What form that precipitation takes—hail, rain, drizzle, freezing rain, sleet, or snow—will depend largely on temperature, both in the atmosphere and on the ground.

The first thing that meteorologists take into account when trying to predict precipitation (in terms of its probability and what form it will assume) is the season. In summers they are on the lookout for thunderstorms, while in winter they are watching for

snow and sleet. In addition, precipitation in summer tends to be concentrated in local areas and very intense, whereas precipitation in winters tends to be more widespread and moderate. Predicting precipitation is complicated by the fact that as temperature changes depending on altitude, snow can change to rain or vice versa.

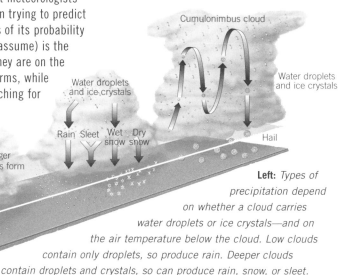

Left: Types of precipitation depend on whether a cloud carries water droplets or ice crystals—and on the air temperature below the cloud. Low clouds contain only droplets, so produce rain. Deeper clouds contain droplets and crystals, so can produce rain, snow, or sleet.

Cumulonimbus cloud

Water droplets and ice crystals

Water droplets and ice crystals

Water droplets

Larger droplets form

Drizzle Rain

Rain Sleet Wet snow Dry snow

Hail

Condensation level

A 100 Percent Chance of Rain?

Meteorologists use what are known as probability forecasts to estimate the likelihood that a given area will get precipitation during the forecast period: a 25 percent or 1 in 4 probability, for example. Probabilities are given in 10 percent increments.

One drawback is that probability forecasts cannot be used to predict when, where, or how much precipitation will occur. Saying that there is a 50 percent chance of rain tomorrow does not mean that it will rain half of the day and nor does it tell you how much it will rain or exactly where. However, when a forecast predicts an 80–100 percent chance of precipitation, it becomes a near certainty. The accuracy of probability forecasts cannot be determined from one such forecast but assessed only by calculating the reliability of thousands. The U.K. Met Office boasts 83 percent accuracy for next-day precipitation forecasts.

In spite of their limitations, probability forecasts do provide useful information on which to base decisions. For example, the launch of a space shuttle or satellite is likely to be scrubbed if a forecast predicts a 60 percent chance of rain that day.

How much moisture is present in the lower levels of the atmosphere is a significant factor: without sufficient moisture, clouds will not form and precipitation will not take place. The amount of moisture in the atmosphere is depicted on a surface map in terms of dew-point temperatures. If there is enough moisture in the air and a cold front is approaching, the chances of precipitation will increase. Fronts serve as a mechanism to lift air up. Rising air is essential for precipitation to occur. As air rises, the moisture it contains begins to cool; then the water vapor condenses to form clouds, setting the stage for precipitation. The water droplets in the clouds continue to grow and at a certain point become too heavy for the cloud to keep suspended— and the result is precipitation.

To make a forecast of precipitation a meteorologist needs to examine current and past weather patterns and assess how these patterns are likely to change in the near future. Such factors as terrain, high- or low-pressure systems, temperature, humidity, and long-term weather statistics must be taken into consideration. Time is also a significant factor in making forecasts: as more data is acquired the more accurate the forecast is likely to be. If meteorologists receive word that the course of a storm has changed based on satellite observations, they will alter their prediction accordingly. Once it is determined that precipitation is likely to occur, the forecaster has to estimate the form it will take: showers, for example, are usually localized, whereas a steady rainfall is more widespread.

The Precipitation Process

Precipitation occurs as a result of the condensation of water vapor as it cools, forming clouds. But before water can return to Earth as rain, snow, or hail, droplets or ice crystals must grow by merging with other droplets or crystals in the cloud—a process called coalescence. Once these droplets or crystals become too heavy for the cloud to hold, they begin their descent.

What is Coalescence?

Precipitation begins with condensation. Relatively warm, moist air rises, often driven upward by a cold front. As the air expands and loses energy, it cools. At that point water vapor begins to condense and form clouds. Then when the water droplets grow too large for the clouds to hold them, they descend as precipitation. The process of growth of these water droplets in a cloud is called coalescence. In stable environments the water droplets tend to remain separate because of air resistance, but turbulence will stir up the droplets and cause them to collide and merge. Now they weigh enough to overcome air resistance and so begin their journey downward as rain. This process generally occurs in clouds where temperatures are above freezing; if the temperatures are below freezing, the supercooled water droplets will merge and freeze into ice crystals. In that case, coalescence involves the collision of water droplets with ice crystals. This is called the Bergeron process. The result is similar: the ice crystals gain more mass and more weight and begin to fall as ice or snow. However, they may change to rain if the ice crystals pass through air where temperatures are above freezing.

Predicting the Intensity of Precipitation

Large-scale, or stratiform, precipitation occurs when the air is slow to rise, especially ahead of warm fronts (see pp. 22–3). Warm fronts tend to produce steady and light or moderate rain or snow in front of them that can last from a few hours to several days. Periods of thunderstorms over a large area are also generated by these fronts.

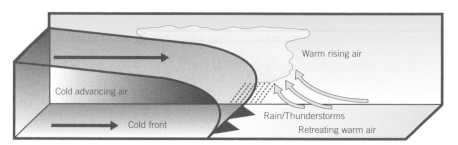

Warm rising air

Cold advancing air

Cold front

Rain/Thunderstorms

Retreating warm air

Above: *As a cold front moves in, cold air flows in under rising warm air, lifting it up higher into the atmosphere, setting in motion a process that can produce intense precipitation.*

Below: *A wind blowing up a mountain produces convective clouds and rain or snow. Having lost its moisture as precipitation, the dry wind is warmed by compression as it rushes down the other side of the mountain. This hot, dry wind, called a Chinook in the Rockies and a föhn in the European Alps, melts the snow in spring.*

Air rises, forming clouds and precipitation

Warm, dry air

Convective precipitation originates from convective clouds: cumulonimbus or cumulus congestus. Convective precipitation varies greatly in its intensity as it falls, and ranges from widespread showers to rain over only a limited locality, as convective clouds have a limited horizontal spread. This form of precipitation is common in the tropics, where warm, moist air rises fast. In mid-latitudes, convective precipitation is associated with cold fronts and often manifests itself as hail or as snow pellets (called graupel).

Precipitation Types

Hail is produced by intense thunderstorms where snow and rain are thrown together in the central updraft of the storm. As they fall, the snowflakes merge with frozen water, resulting in ice pellets. These pellets acquire more mass as more and more supercooled droplets adhere to them. They grow larger as they fall through the cloud. Sometimes the updraft is powerful enough to send them back up to the top of the storm, renewing the process and making the hailstone even larger. Because these hailstones are not exposed to warmer

temperatures in the atmosphere for long enough to melt as they descend, they reach the ground as hail.

Ice storms result from rain that has been supercooled and freezes on impact with cold surfaces: in other words, it is not in the form of ice in the atmosphere.

Freezing rain usually occurs in a narrow band on the cold side of a warm front, where surface temperatures are at or just below freezing. It may coat ground surfaces and be extremely hazardous.

Sleet is defined differently in different parts of the world. According to the U.S. definition, sleet consists of frozen raindrops that bounce upon impact with the ground. In other parts of the world, sleet is a mixture of rain and snow and occurs in a transition region between the two. For more on sleet, see pp. 94–5.

Snowflakes are made up of ice crystals that have merged with one another in their descent. Snow generally occurs when cold arctic air is settling in the area at freezing or below. See also pp. 92–3.

The Anatomy of a Thunderstorm

A thunderstorm can develop only if certain conditions are present: there has to be enough moisture in the atmosphere, ground temperatures must be warmer than the air aloft, and the updrafts of warm air rising from the ground must be strong enough to break through the upper levels of the atmosphere. Thunderstorms acquire their destructive power from two competing forces: sinking cold air and rising warm air.

Thunderstorms can only form when the air is unstable: when the air near the ground is warm while the air aloft is cold. (There is, however, a phenomenon known as thunder snow, which is a thunderstorm in which snow reaches the ground instead of rain.) The temperature difference between the ground and the atmosphere is crucial. Thunderstorms are common in spring because the air aloft retains its winter cold, making the air more unstable.

1 Rising warm air: Thunderstorms need a source of warm air to get started. This can occur when warm, humid air, or an updraft, rises from the ground. It can also occur through advection, when winds blow hot air into the area. In either case, warm air will rise only if it is lighter than the surrounding air. That is why a thunderstorm will occur only when the air aloft is cooler, as warm air is less dense than cold air. Thunderstorms are known as convection storms as they owe their origin to vertical air circulation in which warm air rises and cool air sinks. If the rising air, also known as a thermal, continues to be warmer than the air that it is rising into, it will continue to rise.

2 Moisture: Another essential ingredient of a thunderstorm is moisture. As air cools to its dew point, or the temperature at which condensation occurs, it leads to the formation of clouds. Moisture is what a thunderstorm is made up of: it is a column of water suspended in the atmosphere.

3 A trigger: There is a small stable layer in the atmosphere that can put a break on the convection process, halting the birth of a thunderstorm. That layer—composed of relatively warm air aloft (usually several thousand feet above the ground)—is often but not always the tropopause, the border region between the troposphere and the stratosphere. As air parcels rise into this layer they become cooler than the surrounding air, making it difficult for them to rise further. So something is needed to break open that stable layer and force the rising parcel of warm air through—essentially a trigger. This can occur in a number of ways. The updraft itself might be powerful enough to force the warm, moist air through this layer. A low-pressure system can also lend a helping hand. While updrafts continue feeding warm, humid air into the storm, the storm is simultaneously generating downdrafts of cooler air. It is the action of updrafts and downdrafts of different air temperatures that make thunderstorms so violent.

What Are Thunder and Lightning?

Lightning is the discharge of electricity that can occur either between rain clouds or between a rain cloud and the ground. Lightning represents an attempt to balance the difference in positive and negative charges within a cloud, between two clouds, or between a cloud and the ground. The base of most clouds is negatively charged and it is attracted to a positive charge on Earth. The jury is still out as to why such polarization of static electricity takes place, with some researchers holding that ice formation in the upper levels of the clouds is essential.

The electrical discharge produces a sound wave that is audible as thunder. We see the lightning first because light travels at 186,000 miles (or 300,000 km) per second, about a million times faster than the speed of sound.

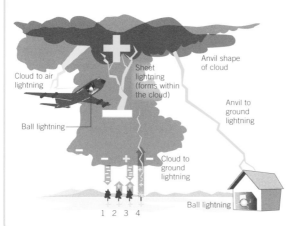

Left: *Cloud-to-ground lightning occurs when a stream of negative charges tries to reach the ground (1) while positive charges from the ground are drawn upward, especially from tall pointed objects (2). The two charges meet (3) and balance themselves out in an upward transfer of positive charges (4).*

4 The thunderstorm: All thunderstorms form within cumulonimbus clouds (see p. 79) and are typically accompanied by heavy rain or hail, lightning, and thunder. Severe thunderstorms are those that, in addition, produce tornadoes (see pp. 32–3), hail that is larger than 3/4 inch (2 cm) in diameter, and winds in excess of 58 mph (93 km/h). The most severe thunderstorms are always pushed ahead of cold fronts. These thunderstorms can be further fueled by jet stream winds. Some thunderstorms consist of two to four thunderstorms: they are called supercells, each of which consists of winds blowing upward in a single rotating updraft.

5 Dying down: Eventually the downdrafts grow in intensity, stifling the warmer updrafts. The source of the storm's energy begins to diminish and the storm soon runs out of steam.

Predicting Thunderstorms

Using sophisticated mathematical models, meteorologists are now capable of predicting the probability of a thunderstorm five days in advance with reasonable accuracy. Applying the same techniques at a more basic level, the amateur weather watcher can also predict the likely occurrence of thunderstorms.

Since thunderstorms can be so dangerous and disruptive, meteorologists have placed a premium on developing ever more accurate methods of predicting them. For the most part they rely on computers to generate conceptual models of meteorological systems that, once combined with observational data from the current state of the atmosphere, can usually lead to reliable forecasts. Computer models have led to a real improvement in forecast services when it comes to thunderstorms. However, meteorologists are not only interested in predicting when or if a thunderstorm will develop at a particular location: these models can allow them to anticipate such factors as how a predicted thunderstorm will move, how severe it will be, and how long it will last.

In general, forecasters try to identify conditions in an environment that will contribute to the formation of a thunderstorm: moisture, instability, and a low-lifting mechanism or trigger (see p. 88). But forecasters also need to view this data in perspective, which necessitates an examination of earlier conditions. They can compare their models with relevant data from observations to gain what they refer to as "ground truth." Forecasters try to envision different scenarios based on such factors as the low-level temperature, wind velocity, and moisture fields. Then they make small changes to each factor in the model to see what effect these changes will have on the overall environment and how these changes will in turn impact on the behavior of the predicted thunderstorm. Because of so many variables this can be an especially challenging task.

Thunderstorms can be predicted with reasonable accuracy up to one week in advance; more precise, detailed forecasts are issued as the day of the predicted event grows nearer. In Australia, for instance, thunderstorm watches are sometimes given 34 hours in advance when severe thunderstorms are possible in a region. However, many thunderstorms develop within hours and with little warning.

Squall Lines

A line of thick cumulonimbus clouds is a sure sign that a frontal thunderstorm is taking place. These prefrontal squall lines are often seen as much as 50–60 miles (80–100 km) ahead of an advancing cold front.

How Far Away is the Storm?

To discover how far away a thunderstorm is, count the time that passes after you see the flash before the clap of thunder. If you wish to estimate the distance in miles, count the seconds between the flash and the thunder and divide by 5. To find out how many kilometers away you are, divide the number of seconds by 3.

Looking at Weather Maps

To predict the chances of a thunderstorm developing using a weather map, look for cold fronts, and dashed lines indicating a trough. A trough is defined as a wind shift line: the wind is blowing one way on one side of the line and a different way on the other. Severe weather is most likely to develop along a trough line or a cold front.

Lines representing occluded fronts in which a cold front has overtaken a warm front (see pp. 24–5) also indicate heavy precipitation. Isobars, which link together areas of equal atmospheric pressure, provide another source of information. A fall in atmospheric pressure indicates a low, which increases the likelihood of storms in the direction the low is headed.

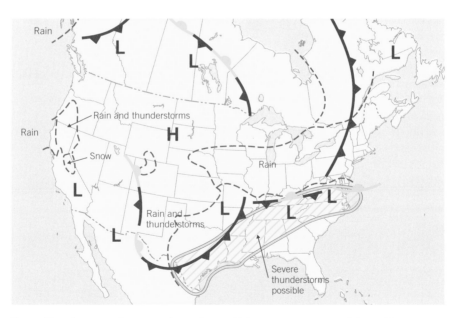

Above: *Thunderstorms frequently form along cold fronts and under conditions of low atmospheric pressure, as seen in this weather map of the United States. Both conditions prevail in a band stretching from Texas to Virginia. The possibility of thunderstorms is further increased because of the presence of warm, moist air, drawn off the Gulf of Mexico.*

Predicting Snow

Even small temperature changes can throw meteorologists' calculations out of kilter when it comes to predicting snow or likely accumulation. In winter, snow may change to rain and then back to snow or descend as sleet or freezing rain, making snow prediction much more problematic than forecasting rain.

Factors to Consider

Given the problems that snow (and other winter weather) can cause, getting it right is important for many people. So snow forecasters have to put a great deal of time into analyzing the data and model charts even on an ordinary winter day. Their objective is to answer several key questions: Will there be precipitation? If so, what form will it take? How intense will it be? Will the temperatures be sufficiently cold for accumulation to occur? How much accumulation can be expected? And how long will the "precipitation event" last?

In addition to accumulation, it is necessary to take into account other factors, such as road conditions. Often accumulation might not be the main concern: an inch or two of snow might be an inconvenience but a coating of freezing rain can present a hazard. Motorists will need to know what kind of road conditions to anticipate before they leave their homes or work, so it is up to meteorologists not only to make predictions but to put their predictions into perspective.

In addition, the probability of snowfall will vary according to season and geographic factors such as latitude and elevation. Close to the equator, for instance, little snowfall is to be expected. Elevation (and the degree of a slope) will also affect whether a snow covering is permanent or not. Snow covering can remain year-round above 16,000 feet (5,000 m) even on mountains near the equator.

Snow tends to fall in narrow bands and in relatively small localities. Transitional zones, or the boundaries between rain and snow, where temperatures change over a small area, can pose particular difficulties for forecasters. Meteorologists also have to consider the velocity of the precipitation. A raindrop reaches the ground much more rapidly than a snowflake. This means that you cannot rely on the position of a cloud

Predicting Snow

Snow is very difficult to predict even for forecasters with satellite and radar imagery and supercomputers at their disposal. Timing is also important: conditions that seem to favor snow may change at the last minute, producing rain or freezing rain instead. The main thing to look for in predicting snow is cold arctic air settling in the area at freezing or below.

Snow Facts

- The highest seasonally cumulative precipitation of snow ever measured was on Mt. Baker in Washington State. Between 1998 and 1999 it received a staggering 95 feet (29 m) of snow.
- A record of almost 187 inches (474 cm) of snow fell in just seven days on Thompson Pass, Alaska, in February 1953.
- The only snow to appear on the equator is at an altitude of 15,387 feet (4,690 m) on the southern slope of Volcán Cayambe in Ecuador.
- Annual snow cover in the northern hemisphere is decreasing at a rate of 3 percent per decade, according to satellite data, a phenomenon attributed to global warming.
- The largest snowflakes on record—15 inches (38 cm) in diameter—were measured in Montana, in 1887.
- About 250,000 avalanches occur each year in the Alps, with France recording more avalanche fatalities from 1993 to 2003 than any other country in the world.

to predict snow, because that cloud might be long gone by the time the snow for which it is responsible actually reaches the ground. Wind can cause snow to blow quite a distance from where it is originally expected to come down. When snow falls from high elevations it can be blown several miles away from the cloud from which it originated. Snowflakes also land at different rates: larger snowflakes land first while smaller snowflakes that have less mass trail behind. That is why you can often experience a snowstorm that begins with a thick, heavy snowfall but over time produces smaller and smaller snowflakes.

Types of Snowfall
All snowflakes are composed of hexagonal ice crystals, but they can fall in a variety of ways and for different durations. When used on its own, the term "snow" refers to precipitation of significant duration and extent. A "blizzard" is a severe storm that lasts three or more hours, bringing with it low temperatures, strong winds, and poor visibility because of blowing snow. A "flurry" or a "snow shower" is a snowfall that suddenly stops and starts and can abruptly change in intensity; accumulation and coverage are limited.

A "snow squall" is characterized by strong winds, flurries, and poor visibility. "Blowing snow" is produced by the wind from the Earth's surface to a height of about 6 feet (2 m). "Drifting snow" is snow that is blown to a height of less than 6 feet (2 m). "Snow pellets" are brittle bits of snow that occur in snow showers when surface temperatures are about 32°F (0°C) and break up upon impact with the ground. "Snow grains" are minute, opaque grains of ice and tend to fall in small quantities.

Predicting Ice and Frost

Ice is a densely packed form of snow. Although commonly associated with winter storms, ice can pummel the ground in summers in the form of hail created by severe thunderstorms. Depending on conditions, ice can melt quickly or linger for decades. Frost is defined as a deposit of minute ice crystals that forms when water vapor converts directly to ice crystals at a temperature below freezing.

Whether precipitation will take the form of ice or snow (or whether snow will turn to ice once it is on the ground) depends on a number of factors, including the snow accumulation rate, the air temperature, and the weight of the snow in the upper layers. Ice can form in as little as a few hours or it can bide its time and take decades. Ice is defined as densely packed material formed from snow that does not contain air bubbles. Ice can occur in summer just as it can in winter.

In summer ice takes the form of hail and is created in the turbulent central updrafts of intense thunderstorms when both rain and snow mix freely. Ice pellets form when water droplets begin to merge and freeze; sometimes just as they reach the bottom of the thunderstorm they are snatched away by the force of the updraft and swept up to the top of the storm, where additional raindrops are frozen onto them. If these ice pellets continue to grow by this constant process of accretion they eventually become hailstones. Hail is simply precipitation that forms lumps of ice, as small as peas or cherries but reaching the size of oranges in some cases. When they are heavy enough and the updrafts can no longer keep them suspended in the cloud, they descend to the ground. But that is

not to say that ice pellets also do not make it to the ground. Ice pellets arrive in translucent frozen raindrops or in snowflakes, but they bounce and make a sound on impact. Exposed to the Sun, they sparkle and produce a diamondlike effect. But their appearance is deceptive: they can also do a lot of damage. There is also an icy phenomenon called snizzle that, as its name suggests, resembles a frozen drizzle. It is made up of extremely small particles of snow and ice that fall lightly.

Sleet

Sleet is often produced by a wintry mix of rain and snow. However, in these regions of transition between different types of precipitation, sleet tends to be brief. A "sleet storm," on the other hand, is another story. In the sleet storm, most of the precipitation is completely sleet although it usually starts out as freezing rain. In these cases the sleet falls for several hours and can produce over 1 inch (2.5 cm) of accumulation.

But what is sleet? That depends on which definition you choose. Sleet is defined differently in the U.K. and the United States. In the U.K. sleet refers to snow that has partially melted in its descent because the atmospheric temperature is warm enough to melt it but not so warm that it turns to rain. In other words, under this definition sleet consists of partially melted droplets, a mixture of snow and rain. Whether it will remain sleet depends on whether the ground temperature is below zero, in which case it will form a treacherous layer of ice called black ice. In the United States, sleet consists of ice pellets (frozen raindrops). In this case the snowflakes begin to melt as they pass through air of warmer temperature, but if they descend through a layer of colder, subfreezing air close to the ground they may freeze all over again, producing ice pellets. Ordinarily they will bounce and leave no trace but they may form ice if they mix with freezing rain—that's where the rain and ground are sufficiently cold that the raindrops freeze on impact. You can observe this effect on a tree where all the branches are covered in a uniform layer of very clear, shiny ice.

Frost

Frost forms directly from water vapor without condensing as dew. Frost is formed by sublimation, the process by which water vapor becomes a solid—namely ice. It occurs when the frost point is below freezing. (The frost point is simply the dew point at freezing temperatures. The dew point is the temperature to which the air must be cooled to reach saturation under constant atmospheric pressure.) Dew seldom forms before frost, but it can happen when the temperature is well above freezing early in the evening and then falls below freezing during the night. Frost that forms from liquid water when temperatures drop below zero is often referred to as frozen dew.

In colder months, frost develops as the air temperature decreases to below the frost point. Much of the water vapor that turns into frost (or dew) actually comes from evaporation from the soil and transpiration from plants and grass. When the soil and vegetation is very dry, little frost (or dew) will occur. Frost tends to form first above the ground—on rooftops or on automobile hoods—because the air above ground is slightly cooler than it is on the ground.

Ice Fog and Freezing Fog

Ice fog is a relatively rare phenomenon that typically occurs only in regions where temperatures are well below freezing such as the Arctic or Antarctic, usually in clear, calm weather. Ice fog is any kind of fog that is frozen—its water droplets are suspended in the air in the form of minute crystals of ice. Ice fog is seldom seen at temperatures warmer than -22°F (-30°C), while it is almost always present at temperatures of -50°F (-45°C) in the vicinity of a source of water vapor, such as a sea or fast-flowing streams. Ice fog should be distinguished from freezing fog, which occurs when the water droplets composing the fog freeze on surfaces, forming white rime ice, as seen on mountaintops exposed to low clouds.

Using Atmospheric Pressure and Winds

Winds are formed by a tug of war between regions of high and low pressure; as warm air in low-pressure regions rises, colder air from high-pressure regions blows in to fill the void. Most major changes in weather are due to a combination of air-pressure differences and the resulting winds they create.

Atmospheric pressure is defined as the weight of air above any area in the Earth's atmosphere (for a full definition of air pressure and its measurement, see pp. 14–15). Atmospheric pressure influences the behavior of air masses, resulting in the creation of areas of high pressure, known as anticyclones, and low pressure, known as depressions. Rising atmospheric pressure indicates the approach of fair weather, and falling atmospheric pressure warns of storm systems. Air pressure can decrease, in two ways: the mass of air above an area may decrease or a mass of air may rise. When the mass of air decreases in the upper level of the atmosphere, which occurs in a thunderstorm, it is called an upper-level divergence.

Troughs and Ridges

Fronts force the air to rise, causing the surface pressure to decrease in the vicinity of the front, creating a trough. Cold fronts have a more defined pressure trough than warm fronts. Troughs are represented on weather maps by a dashed line. These areas of relatively low pressure are unstable and tend to have high moisture content; as a result they are a source of precipitation, especially thunderstorms. Troughs tend to form in the warmer months.

Troughs may intensify rapidly, with strengthening winds that lift warmer air. The more quickly the air rises, the lower the pressure becomes.

By contrast, ridges are regions of high pressure and are generally characterized by stable conditions, although they are associated with winds that bring coastal showers in advance of the ridge. In zones where troughs and ridges interact, the atmosphere can be unstable and produce storms or rainfall.

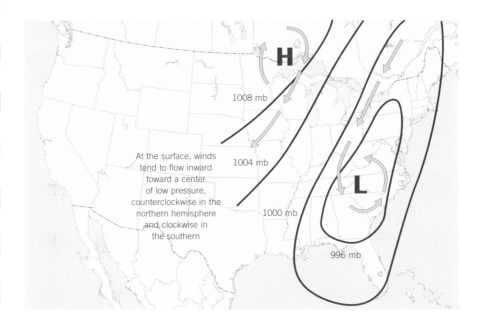

Above: *In the northern hemisphere, winds flow counterclockwise around and inward toward a low-pressure area at increasing speed. The wind circulation forces the air upward, carrying heat and moisture up into the atmosphere, which results in clouds and precipitation. In the southern hemisphere, wind blowing into a low-pressure system moves in a clockwise direction.*

The difference in air pressure between regions in the atmosphere causes air to move. Air moves from a region where it is denser (high pressure) than the surrounding area to one where it is lower (low pressure). As winds blow into a low-pressure area, the air can be uplifted.

Warmer, humid air is less dense than cooler air, so it will rise; cooler air sinks. The former is an updraft; the latter a downdraft. Both of these are involved in the formation of a thunderstorm. The updraft becomes a region of relative low pressure. Meteorologists call this a mesolow. (Mesolows and mesohighs are measurements on the mesoscale, which relates to such meteorological phenomena as wind circulation, cloud patterns,

thunderstorms, and squall lines.) Low-pressure areas swirl counterclockwise in the northern hemisphere and clockwise in the southern hemisphere, sucking in air that quickly rises. As it rises, it begins to cool at a uniform rate. The cooling process in turn sets in motion condensation of water vapor, which leads to cloud formation and rain. The downdraft, containing rain-cooled air of higher density, is going in the opposite direction, accelerating toward the surface. The downdraft is associated with a high-pressure area, called a mesohigh. This will cause the surface pressure of the area to rise in the presence of a thunderstorm. That same process can cause a decrease in pressure at upper levels of the atmosphere as denser air descends.

Stable and Unstable Air

Atmospheric stability depends on whether an updraft of warm air will lose its buoyancy, cool, and sink. If it continues to rise and break through higher layers of the atmosphere, it can produce unstable conditions, resulting in violent thunderstorms.

When meteorologists use the term "stability" in reference to the atmosphere, they are being quite literal. When the air is stable, the weather is generally calm. Stable air does not rule out the possibility of rain or snow, but under stable conditions precipitation tends to be slow and steady and cover a large area. Violent thunderstorms and intense snowfalls occur only when atmospheric conditions are unstable. Air stability also influences cloud formation and affects the type of cloud that will emerge.

What Causes Stable or Unstable Conditions?

Whether air is stable or unstable depends on the temperature of rising air relative to the temperature of the surrounding air that it is passing through. The temperature of the stationary air will vary from one location to another. Packets of air rise when the air is heated. As air rises and expands, the atmospheric pressure closer to the surface begins to fall off. By the same token, the temperature of the rising packet begins to cool. The air packet cools at a rate of 5.5°F (3°C) for each 1,000 feet (300 m) it rises. It cools at this rate as long as the humidity in the air is not condensing, meaning that no clouds are forming. The rate of cooling remains the same regardless of the temperature of the surrounding air. The cooling effect results from expansion as the air moves into an area of lower pressure. (Similarly, as colder air sinks, it warms at the same rate—it is descending

into an area of higher pressure—because it is being compressed, or the air molecules are being squeezed together.)

If the process continues and the air continues to cool, it will eventually reach the same temperature as the air around it: when this happens, it will start to sink. This brings about stable conditions. Sometimes, though, rather than sinking, the air packet may continue to rise. This occurs when the air stays warmer than the surrounding air. This creates unstable conditions. On a very unstable day, air can rise to 50,000 feet (15,000 m) or higher, which can cause intense thunderstorms.

Let's take an example. On a spring day of stable air conditions, meteorologists might record a temperature of 70°F (21°C) on the ground, 67°F (19°C) at 1,000 feet (300 m) above the ground, and 63°F (17°C) at 2,000 feet (600 m), based on data received from weather balloons. Then a cold front forces warmer air to rise from the ground. When the 70°F (21°C) air packet reaches 1,000 feet (300 m) it has cooled to 64.5°F (18°C) degrees and is cooler than the surrounding air, which is 67°F (19°C). If nothing else changes, that air packet will sink, so we can expect stable conditions. But let's say that a cold front moves in aloft, while the temperature on the ground remains the same at 70°F (21°C). At 1,000 and 2,000 feet (300 and 600 m), the air has turned cooler. Now when the packet of 70°F (21°C) air rises—cooling 5.5°F (3°C) for each 1,000 feet (300 m)—it is warmer than the surrounding air and continues to rise. Unstable conditions will result. Unstable conditions could also result if the air on the ground warmed significantly while the air at 1,000 and 2,000 feet (300 and 600 m) stayed the same.

If we factor in humidity, the air will become even more unstable. That is because, as the air cools as it rises, the water molecules that it contains start to slow down. This slow-down causes the molecules to change from vapor to liquid or ice, while the kinetic energy of their movement changes into another form of energy—heat. This "latent heat" warms the surrounding air and causes yet more instability. That's one of the reasons why thunderstorms frequently occur on humid summer days.

Cloud Formation in Stable and Unstable Conditions

The cloud-forming process gets under way as air begins to cool, becomes saturated, and then condenses. The air within the cloud is either stable or unstable, and the cloud's structure will depend on the degree of stability or instability. Stable air resists convection—the process that allows clouds to grow vertically—while unstable air does the opposite. As a result, clouds with stable air develop horizontally, forming stratus clouds, which are sheetlike in appearance. Convection clouds, however, will build up into cumulus clouds that resemble towers distinguished by billowy shapes. Clouds that show such vertical growth vary from fair weather cumulus to cumulonimbus (see pp. 78–9).

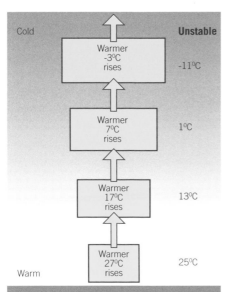

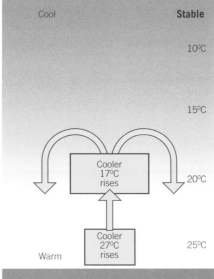

Above: *Under stable conditions, warm air rises only so far before it reaches the same temperature as the air around it and begins to sink. This makes cloud formation and precipitation less likely. Under unstable conditions, updrafts of warm air continue to rise and thunderstorms can be born.*

Wind Speed and Direction

Which way the wind is blowing is less important than where the wind is coming from—that is where tomorrow's weather in your locale is happening today. How fast the wind is blowing, whether it is a light breeze or a category 3 hurricane-force wind, will also play a crucial role in what kind of weather you can anticipate.

The wind acts as a mechanism to compensate for inconsistent temperatures on the Earth's surface and to equalize the pressure differences due to those differences in temperature (see pp. 18–19). The greater the differences in pressure, the stronger the wind will be. The movement of air from a high-pressure area into a lower-pressure area is called convection.

Wind speed is the speed of air moving relative to a fixed point on Earth. When winds reach very high speeds they can become hurricanes, typhoons, or gales. Wind speeds increase at higher elevations where the air is thinner and there is less frictional drag. At the surface, the strongest winds are usually found near cold fronts and low-pressure systems. Conversely, winds are normally light near high-pressure systems.

The most famous scale developed to measure wind speed, known as the Beaufort Scale, was developed in the early 19th century by Britain's Admiral Sir Francis Beaufort (1774–1857) to help sailors determine wind speed on the basis of visual observations of wave heights. The scale is still in use today, a tribute to its usefulness. Over land, wind speed is given in miles or kilometers per hour; over water it is given in knots. A knot is exactly equal to 1/60th of a degree of latitude or a minute of latitude per hour. A knot also translates to 1.15 mph (1.85 km/h).

When forecasters talk about wind direction they are referring to the direction from which the wind has come, not the direction in which it is blowing. Direction can also be referred to in terms of upwind and downwind—if you are moving upwind, you are moving against the wind.

What Wind Can Tell Us

Wind can tell forecasters many things. Wind speed and direction will give you a good understanding of what to expect, as the weather upwind is coming your way. Wind is an indication of two major weather developments: temperature and moisture advection. In meteorology, advection refers to the horizontal transport of an atmospheric property (like heat or moisture) by the wind. If temperatures are warmer upwind, then it would indicate that the wind will bring warmer temperatures downwind, and vice versa. The same holds true for humidity.

When winds are light, all things being equal, temperatures will be cooler at night and warmer during the day. During the day, heat builds up at the surface; if the winds are light they do not have sufficient force to mix the heated air with the colder air aloft, allowing heat to build up. At night, if winds remain light, the surface loses (or radiates) heat, making it colder at the surface. If wind speed is stronger during the day, the wind will disperse the warmer air at the surface, resulting in cooler temperatures.

The Beaufort Scale

Force	Mph (km/h)	Description
0	0–1 (0–1.5)	Calm
1	1–3 (1.6–6)	Light air
2	4–7 (7–12)	Light breeze
3	8–12 (13–20)	Gentle breeze
4	13–18 (21–29)	Moderate breeze
5	19–24 (30–39)	Fresh breeze
6	25–31 (40–50)	Strong breeze
7	32–38 (51–61)	Near gale
8	39–46 (62–74)	Gale
9	47–54 (75–87)	Severe gale
10	55–63 (88–101)	Storm
11	64–72 (102–116)	Violent storm
12	73–83 (117–133)	Hurricane

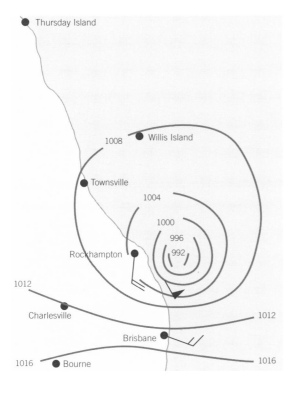

Left: *The arrows on weather charts indicate the forecast wind speed and direction. When winds swirl around a low in the southern hemisphere, as depicted here for Australia, they blow in a clockwise direction. They blow in a counterclockwise direction around a high. (It is the opposite in the northern hemisphere.) Winds will typically blow along the line of isobars—the lines connecting areas of equal pressure—but turn toward a low and away from a high. The numbers on this chart indicate the barometric pressure. The closer together the isobars, the stronger the winds will be.*

Local Winds

In addition to the global circulation of winds—the westerlies and polar easterlies, for example—winds affect weather on a more local scale. Local winds are defined as those that occur in a small region.

Land and sea breezes, and mountain and valley breezes, are all local winds caused by fluctuations in temperature and pressure over irregular terrain. Most local winds typically produce a particular type of weather. Many local winds, such as the Sirocco or Khamsin, are given names based on their particular geographical, topographical, or seasonal characteristics.

Land and sea breezes are local winds that usually blow for a distance of about 30 miles (50 km) on and off the shore. Variations in pressure typically cause breezes to blow landward during the day and seaward at night. In summer, the land is warmer than the sea by day and colder than the sea by night. The warm air rises, and the variations in pressure create a system of breezes.

Mountain and valley winds are caused by differences in air pressure and temperature in mountainous regions. In the summer, the sheltered air along the bottom of a valley heats up. The warm air rises up the mountains, while the cooler air above the

middle of the valley sinks. Warm air tends to blow up the mountains in morning and up the river valleys in the afternoon. These are called valley winds. During the evening, as the air chills, the process reverses as cooler, denser air moves down the slopes as a mountain wind.

The Mistral (meaning "master wind" in Provençale) usually develops as a cold front moves down through France, pushing up the air in the Alps before rushing down the Rhone Valley at up to 80 mph (130 km/h). Its effect is felt mostly in the winter and spring, with a cold, strong northwesterly wind along the Mediterranean coast of France. It is a katabatic wind—a wind caused by air cooled over the mountains by a high-pressure system.

The Vendavales (meaning "gales" in Spanish) are created by a low-pressure system that enters the western Mediterranean from the Atlantic in late autumn through winter. This causes strong southwesterly winds to blow in the Straits of Gibraltar and brings thunderstorms.

Right: On sunny days, the land warms faster than the sea. The warmed air above the land rises and draws in cooler sea air to replace it, called a sea breeze. At a height of about 3,300 feet (1,000 m), the air above the land blows out to sea. During late evening, the flows are reversed because the sea keeps its warmth longer than the land. Air flowing away from the shore is called a land breeze.

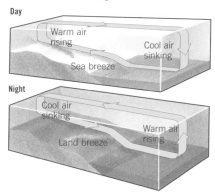

Day

Warm air rising

Cool air sinking

Sea breeze

Night

Cool air sinking

Warm air rising

Land breeze

Wind Chill Index

The wind chill index will be familiar to anyone who tunes into a weather forecast in winter. The index is meant to give a relative sense of how cold it will feel to exposed skin outside. All things being equal, the colder the wind chill index, the colder it will feel. The index takes into account not only temperature but wind conditions. A strong wind will make it feel much colder even if the temperature of the air is relatively moderate. But the index has some limitations: winds can change in intensity and their speed is not constant. Moreover, the index does not incorporate other factors, such as humidity or sunlight. And how a person will respond to cold weather will always be very subjective.

The Sirocco (from the Arabic sharq, meaning "east") is also produced by a depression, in the central Mediterranean. The low causes a strong southerly wind to blow out of North Africa, carrying a great deal of dust and sand. As the Sirocco moves across the Mediterranean and into southern Europe, it also picks up moisture. These winds are most common during the autumn and the spring.

The Khamsin (meaning "fifty" in Arabic, as it can last about fifty days) is caused by depressions moving eastward along the southern Mediterranean, or along the North African coast, from February to June. Ahead of these low-pressure areas the wind blows from the southeast off the Sahara Desert. The wind's temperature can be 104°F (40°C) with very low humidity. As this wind blows across Libya, Egypt, and as far east as Saudi Arabia, it dries up everything in its path and fills the air with sand.

The Föhn (from the German for "west wind") is a mountain wind that commonly occurs in the Alps, but the term can be applied to any warm, dry wind blowing down into the valleys from a mountain. Föhn winds are produced when wind is forced over a mountain range; it cools as it moves upslope, producing precipitation and removing moisture from the air. As the dry air descends on the other side of the mountain, it gets warmer as it comes under greater atmospheric pressure. The result is strong, warm, and dry winds.

Chinooks (named after the Chinook Native Americans) are a variety of Föhn wind observed in the west of North America, where the Canadian Prairies and Great Plains end and the mountains begin. Chinook winds can raise winter temperatures by as much as 70°F (40°C).

Santa Ana winds are a type of Föhn wind, the result of air pressure build-up in the high-altitude Great Basin between the Sierra Nevada and the Rocky Mountains, in the United States. In autumn and early winter, this air mass spills out into the surrounding lowlands, creating strong, warm, and dry winds.

Making Use of Your Resources

Now that you have an understanding of weather and its causes, it's time to move on to the next step: making your own weather forecasts. Your forecasts will be all the more accurate if you are able to make full use of all your resources, which may include your own weather instruments, weather maps, your observations of the clouds, and even some old-fashioned weather lore.

Meteorological Scales

First you should be familiar with the different scales meteorologists use to analyze weather conditions:

Synoptic scale: This is large-scale weather: fronts and pressure systems.

Mesoscale: This is local weather, which takes into account the effects of topography, bodies of water, the proximity of cities, and so on.

Vertical scale: This is the scale that looks at how atmospheric conditions—areas of high and low pressure, updrafts and downdrafts—impact on weather at the surface.

Know Your Local Climate

Make sure you have some familiarity with average weather conditions in the place for which you are forecasting. Finding out the averages of temperature, precipitation, and so on over an extended period will give you the climatology of your location. Although exceptional conditions can occur, most

readings will fall within the normal range based on historical measurements. In fact, the Persistence Method (also known as the Climatology Method) of prediction uses precedent as a basis for forecasting, although this method is limited in areas that experience frequent changes in weather conditions.

Tips for Forecasts

1 Look at the current weather conditions, as shown on Internet or TV weather maps. Pay particular attention to any significant weather events such as low-pressure centers or warm and cold fronts that could influence the weather in your area.

2 Use radar or satellite images upwind (usually west) of your location from 12 hours ago to find out about the movement of air masses, lows, and highs, and then compare those images with more recent ones to see how quickly these systems have moved. Estimate how these systems might affect the forecast area.

3 How much precipitation you are likely to expect will depend first of all on what season it is. In summer, precipitation often

comes in the form of thunderstorms—brief, intense interludes of rain over concentrated areas—whereas in winter precipitation may take a variety of forms, such as snow, sleet, freezing rain, and hail. Snowstorms are more often light to moderate and cover a fairly widespread area. That makes forecasting precipitation in winter trickier because slight variations in temperature (in the atmosphere or on the ground) can change snow to rain or vice versa. That will have an effect on accumulation, too, since 10 inches (25 cm) of snow equals about 1 inch (2.5 cm) of water.

4 A few rules are worth keeping in mind when predicting temperature: the lowest temperature of the day usually occurs in the morning and the highest in the afternoons. Light winds and clear skies will cause greater heating of the surface. When you are trying to determine temperatures for the following day, look at temperatures upwind (usually west) of where you are, and bear in mind the cloud cover and wind speed. A chance of precipitation will also have a great bearing on the temperature. An error in judgment on even one of these factors may cause your forecasted temperature to be off by as much as 20°F (10°C).

5 The type and number of clouds in the sky can tell you a lot about the weather to come (see pp. 70–83). To recap, thickening cirrus clouds signal precipitation within 12–24 hours. An overcast sky with stratocumulus clouds is a sign that showers will develop. Puffy cumulus clouds mean that a fair day is in store, while the appearance of cumulonimbus means that thunderstorms are imminent.

6 Once you determine the barometric pressure and measure wind speed and direction you will be in a position to forecast the weather that is most likely to occur in the next 24–48 hour period. For instance, if the wind speed is steady, while the sea-level barometric pressure is 30.10–30.20 inches you would expect fair weather with little temperature change over the next couple of days. But an increasing wind with a barometric pressure reading that is dropping rapidly indicates strengthening wind and rain in the next 12 hours.

7 And finally: try, try, and try again! Even professional forecasters do not get it right all the time. Don't allow yourself to become frustrated. Continue to make forecasts, using what you have learned from your previous forecasts, including your mistakes. Try forecasting in different weather conditions so that you can hone your skills and increase your accuracy.

Weather Forecasting Checklist

For beginners:
- ☐ Barometer
- ☐ Wind vane
- ☐ Newspaper and Internet weather maps
- ☐ Cloud-watching observations
- ☐ Logbook

For advanced weather watchers:
- ☐ Stevenson shelter
- ☐ Thermometer
- ☐ Hygrometer
- ☐ Anemometer
- ☐ Rain gauge
- ☐ Weather software

Using Weather Maps and the Internet

Weather maps can be found in newspapers, on the Internet, and on television. Familiarize yourself with the range of maps available, their potential uses, and the symbols employed. The Internet is also an increasingly valuable resource for the amateur forecaster.

Weather maps depicting current conditions can give you a good picture of how atmospheric conditions are likely to change over the next 24-hour period. A weather map will allow you to estimate how fast air masses are moving and to track warm and cold fronts, and high- and low-pressure systems.

To be able to read a weather map, you will need to acquaint yourself with the standard notation of symbols and numbers (see pp. 52–3). Different numbers and symbols represent various types of data which will not only indicate current weather conditions but will also provide you with information about the positions of significant meteorological features such as fronts, cyclones, and anticyclones that are likely to affect your area in the near future. What are known as surface observation symbols will tell you the temperature, dew-point temperature, cloud cover, as well as pressure and wind observations.

Types of Maps

Humidity maps: Look ahead of a front: that is where you will find a region of high humidity. Low humidity should be located behind a front and near highs. Regions of high humidity are likely to produce precipitation, especially thunderstorms. The higher moisture content of the air provides fuel for these storms. Air masses that are formed over or close to bodies of water will also bring a great deal of moisture.

Precipitation maps: These maps are one of the best guides to forecasting future precipitation in your locale. They rely on different types of images from Doppler radar, which can detect the location and intensity of storms (reflectivity), the speed and direction of wind (velocity), and the total accumulation of rainfall (storm total). A reflectivity image indicates the location of rain, snow, or other precipitation and how intensely it is falling. Meteorologists use a corresponding color key to interpret the reflectivity image. Each color represents a different level of intensity. Look for areas where precipitation is occurring and make note of its intensity.

Pressure maps: Look for areas of low pressure. They are usually close to a front and are likely to bring precipitation. High-pressure areas are generally not associated with precipitation. A rapid change in pressure over a short distance will produce strong winds.

Forecast high and low temperature maps: Use these maps, based on computer models, as a reference to compare with your own forecasts. These models by themselves cannot predict weather but they are helpful as guides.

Upper-air maps: These depict conditions in the atmosphere between 1 and 10 miles (1.6–16 km) above the ground. The data incorporated into these maps is mainly taken from weather balloons which measure upper air conditions over a particular location. Such maps will indicate the position of winds and jet streams; the positions of short waves, which produce cloudiness and precipitation; and advection—the horizontal movement of air—which will effect changes in moisture and temperature.

Using Satellite Imagery

Look for the latest weather satellite photos on the Internet to find out the location of clouds, which can indicate the presence of fronts and low-pressure systems.

The scale of most satellite images is limited; the resolution is typically a surface area from 3 x 3 miles (5 x 5 km) up to 12 x 12 miles (20 x 20 km) for each point on the image. As a result, only dominant weather features are likely to show up. Generally, two types of satellite images are used: Visible (VIS) and Infrared (IRI). The VIS image is similar to a normal black and white image of the Earth. It depicts cloud cover, with the higher, thinner clouds being duller, and the lower, more reflective cloud layers being whiter. These images are limited because they can only be recorded in daylight and they are seldom used by the media.

Far more common are the IRI images; these are made from temperatures readings of each point observed. These readings are depicted in various shades of gray, with the lowest temperatures (which occur at the highest elevations, typically high clouds) shown in white. Lower clouds, fog, and warm oceans or cooler surface temperatures show up in dark gray. The hottest areas on the ground will appear black. These images, unlike the VIS images, are recorded 24 hours a day. However, IRI images do not directly show storms, rain, cloud, or cloud-free regions, which can only be inferred from the temperature differences.

VIS images can supplement IRI images by distinguishing between features that show up on the IRI image as similar shades. Even so, only those areas closest to the satellite will show up, so that higher features such as clouds can sometimes obscure important weather features below. That is why observations from weather balloons, airplanes, and weather stations on the ground are necessary in order to gain a more complete picture of meteorological activity.

Using the Internet

The Internet offers forecasts that are constantly being updated, and if you are not satisfied with just a forecast, you have the option of consulting Doppler radar images, satellite images, and a range of weather maps. Newspapers are increasingly putting more meteorological data on their websites and the web has the advantage that space is not as limited as it is in print editions.

For official weather service information, try the National Weather Service in the United States, the British Meteorological Office, or the Australian Meteorology Service. In addition to official weather services, you can find any number of private meteorological services, including Accuweather, the Weather Underground, the Weather Channel (also on cable TV), and Intellicast. For a full listing of useful websites, see p. 221.

Folklore Forecasting: Nature

Long before the invention of the thermometer and barometer, never mind weather satellites and supercomputers, humans were preoccupied by weather. That's understandable since their very survival depended on it. Farmers, shepherds, fishermen, and sailors learned to look for signs in nature to predict what was in store for them. Much of this age-old wisdom can still be of great use to the weather tracker.

There is almost nowhere in the world where you cannot find a tradition of weather folklore. In spite of all the cutting-edge technological innovations adopted by weather forecasters, millions of people still adhere to traditional methods of foretelling weather by relying on nature. Some weather folklore has long since been discredited as superstition, but a surprising number of weather proverbs, sayings, and rhymes actually turn out to have some validity. People have been using the scientific method—that is compiling observations, reaching hypotheses, and testing them— long before they had a name for it. Bear in mind that most of these observations were made of local conditions, as without technology people had no way to understand how conditions elsewhere affected the weather in a particular location.

1 Red sky at night, shepherd's delight. Red sky in morning, shepherd's warning.
Weather systems typically move from west to east. At sunrise and sunset the Sun's rays have to penetrate thicker layers of atmosphere than at other times of the day because of the Sun's low angle. (The atmosphere absorbs shorter wavelengths—the greens, blues, and violets—of the visible spectrum.) Red skies in the morning come from the Sun's lighting the undersides of moisture-laden clouds moving in from the west, which means that precipitation may be possible. Red skies in the evening, however, indicate that the westerly sunlight has an unimpeded path in order to light up moisture-bearing clouds moving off to the east, which might indicate a clear day tomorrow. There are many variations on this theme. For example, in Matthew 16:3 it says: "When it is evening, ye say, 'It will be fair weather: for the sky is red.' And in the morning, 'It will be foul weather today: for the sky is red and lowring.'"

2 When leaves show their backs, it will rain.
Leaves on growing trees will form a pattern which depends on the prevailing wind. Since storm winds tend to blow in the opposite direction from the prevailing winds, the leaves are ruffled backward, exposing their light undersides.

3 Pimpernel, pimpernel, tell me true. Whether the weather be fine or no. No heart can think, no tongue can tell. The virtues of the pimpernel.
Whatever the virtues of the bog pimpernel, it will close when the atmosphere reaches about 80 percent

humidity, a circumstance that makes rain more likely, and open up when it is sunny.

4 When rain comes before the wind, dories, gear, and vessel mind. When wind comes before the rain, soon you'll make the set again.
When rain precedes wind, it is often the result of an approaching front, which can produce unsettled weather for a day or two, so it would be a good idea for sailors to take precautions. Wind before rain, on the other hand, indicates a localized rainstorm, which will soon be gone.

5 When the wind is blowing in the north, no fisherman should set forth. When the wind is blowing in the east, 'tis not fit for man nor beast. When the wind is blowing in the south, it brings the food over the fish's mouth. When the wind is blowing in the west, that is when the fishing's best!
This description of wind direction accurately depicts what happens in areas of low pressure. Easterly winds typically pick up, often producing hot, dry, gusty winds in the summer and bitterly cold winds in winter. Northerly winds, which follow around a low, are cold and blustery, posing a particular challenge for anyone navigating in waters where they blow. Southerly winds bring warm temperatures, which make for congenial conditions to do a bit of fishing. But no wind is more favorable for sailors than a steady westerly wind, which promises fair weather and good sailing for several days running. However, conditions will vary depending on whether the low passes to the north of the observer or to the south,

but to take all the possibilities into account would no doubt require several more stanzas.

6 No weather is ill, if the wind be still.
Calm conditions, especially when accompanied by clear skies, certainly can indicate the presence of a high-pressure system associated with fair weather. But a calm does not necessarily portend clear skies for long, as anyone familiar with the phrase "the calm before the storm" knows all too well. Calm can also prevail when a large thunderstorm cell off to the west may be sucking in a westerly wind into an updraft before it strikes. In winter, calm air and clear skies may indicate an Arctic high-pressure air mass, which can produce frigid temperatures.

7 Rain long foretold, long last. Short notice, soon past.
When a warm front is moving into the area, bringing precipitation with it, there will be cloud cover overhead for some hours before rain begins to fall. A cold front storm is faster moving and not so easily foretold. Its clouds arrive quickly and the rainfall is heavier, but seldom lasts long.

8 When smoke rises straight into the air, fair weather is coming. When smoke hangs low, rain is on the way.
There is some science behind this belief. When a high-pressure cool air mass is approaching, it brings fair weather. Hot air will rise above the cool air, and so will the smoke. A low-pressure system will bring warmer air that hugs the ground, preventing the smoke from rising so that it tends to lie low.

Folklore Forecasting: Animals

Your pet may not be as reliable a forecaster as your local meteorologist, but there is some evidence that animals react to certain future weather patterns before humans do. Dogs and cats, for instance, are thought to sense when a tornado is about to occur, and squirrels seem to foretell cold winters by collecting more nuts than usual. Such animal behavior has become enshrined in traditional folklore.

1 If cats lick themselves, fair weather.
During fair weather, when the relative humidity is low, electrostatic charges (static electricity) can build up on a cat as it touches other objects. Cat hair loses electrons easily, so cats become positively charged. When a cat licks itself, the moisture makes its fur more conductive, so the charge can "leak" off the cat. Many cats don't like to be petted during cold winter weather when the humidity is low because sufficient charge builds up to cause small sparks, which irritate them (and the person who is petting them). Dog-lovers can call on their own canine tradition, which states that when dogs rub themselves in winter, it will soon thaw.

2 Flies bite more before it rains.
High temperatures cause people to sweat more, and a decrease in atmospheric pressure may result in more body odor. Both sweat and body odor attract flies to make a feast of human flesh. However, it would be a mistake to rely on flies as reliable predictors of approaching rain.

3 Crickets chirp faster when it's warm and slower when it's cold.
Crickets can indeed serve as thermometers. Tradition says that if you count the cricket's chirps for 14 seconds and then add 40 you will obtain the temperature in Fahrenheit at the cricket's location.

4 A cow with its tail to the west, makes weather the best.
A cow with its tail to the east, makes weather the least.
Animals tend to graze with their tails into the wind as a precaution. This way they will pick up the scent of a predator approaching behind them as it is blown by the wind. In the mid-latitudes, easterly winds often bring rain and west winds often bring fair weather, so a grazing cow's tail can act as an indicator of weather.

5 Seagull, seagull, sit on the sand.
It's a sign of rain when you are at hand.
This saying is based on the observation that birds tend to roost more during low-pressure conditions than during high pressure. One explanation may be that flying might be more difficult in low pressure because of the diminished density of the air.

6 If on February 2nd (Groundhog Day) the groundhog sees its shadow, six weeks of winter remain. If not, spring will follow immediately.
Of all the creatures that are believed to predict weather patterns, there is one that has achieved more fame than any other—

Punxsutawney Phil, the groundhog that makes his appearance in the Philadelphia town of that name every February 2. The first official Groundhog Day took place on February 2, 1886, with a proclamation in the Punxsutawney Spirit: "Today is groundhog day and up to the time of going to press the beast has not seen its shadow." The groundhog was given the name "Punxsutawney Phil, Seer of Seers, Sage of Sages, Prognosticator of Prognosticators, and Weather Prophet Extraordinary." His hometown was called the "Weather Capital of the World." If the groundhog saw no shadow, spring would arrive early; but if it noticed its shadow and scampered back in fear inside his burrow, winter would last for another six weeks. Unfortunately, his predictions are said to be no better than 50 percent accurate—no better than chance.

7 The narrower the stripe on a wooly bear caterpillar, the colder and longer the winter.

Wooly bear caterpillars are dark and hairy in appearance, and curl up into a ball when touched; they are black at both ends, with a reddish-brown band in the middle. The use of these caterpillars (which are the pupae of the tiger moth, Isia isabella) dates back to observations in 1608 by naturalist Edward Topsell, who noted that they tended to roam a good deal and survived later in the fall than other caterpillars. They were known to Topsell's contemporaries as "bear worms" because of their un-caterpillar-like tendency to hibernate. Today, these caterpillars are commonly referred to as "wooly bears." According to tradition, the narrower the reddish-brown band, the colder and longer the winter, and the wider the band the

milder the winter. The width of the band is thought to forecast "average" temperatures for the entire winter, not particular cold spells or storms, a failing that became vividly apparent when the wooly bear predicted a mild winter for 1888. That was the year of the Great Blizzard (also known as the White Hurricane), which raged for four days in March, burying the entire east coast of the United States in snow and killing more than 400 people. However, one experiment showed that the color of the caterpillars' bands had more to do with whether they grew up in a wet or dry area. Another study connected the number of brown hairs to the caterpillar's age, which relates to the previous spring's climate, not the following winter's. Whatever the truth, the caterpillar's ability to predict winter weather seems to be nil.

8 If the squirrels are burying lots of nuts, we'll have a hard winter.

Squirrels have been ruled out as forecasters following numerous studies. If they bury an abundance of nuts, naturalists say, it isn't a sign that they're storing away food for a long winter but rather that they are taking advantage of a good crop of nuts.

9 If birds and bats fly low to the ground, rain is coming.

Birds and bats prefer to fly when the air is denser, which gives them added lift. Denser, drier air also allows birds and bats to fly at higher altitudes. Air is denser during conditions of high pressure, which are usually accompanied by fair weather. When the air is thinner (less dense)— conditions that can portend rain—birds will fly much lower to the ground.

THE WEATHER LOGBOOK

Now that you know what to look for and have assembled a weather station, including a thermometer, anemometer, hygrometer, and barometer, you can begin making observations. In this logbook you can record a variety of weather conditions, such as temperature, wind speed and direction, relative humidity, atmospheric pressure, and cloud types. Most authorities recommend making—and recording—observations twice a day, preferably in the morning and evening. When there are special weather events, such as high winds or heavy storms, you will undoubtedly want to monitor the weather more frequently.

This logbook also allows you the opportunity to make your own forecasts and then go back and see how accurate they were. You will almost certainly make mistakes—even professional meteorologists are not correct 100 percent of the time. By making these routine observations you will become more aware of weather patterns over the weeks and months, which will allow you to refine your forecasts. You can test your skills further by studying and predicting weather conditions in other locations. At some point you might want to share your observations and comments online with other amateurs. But to do that you will need the discipline to maintain your logbook regularly.

Monday

Date
and Time .... Location ....

Relative Humidity ....   Wind Speed and Direction ....

Atmospheric Pressure ....   Cloud Cover..... oktas/overcast   Temperature ....

Cloud Types ....

General Observations ....

Predictions for Tomorrow ....

Tuesday

Date
and Time .... Location ....

Relative Humidity ....   Wind Speed and Direction ....

Atmospheric Pressure ....   Cloud Cover..... oktas/overcast   Temperature ....

Cloud Types ....

General Observations ....

Predictions for Tomorrow ....

Wednesday

Date
and Time .... Location ....

Relative Humidity ....   Wind Speed and Direction ....

Atmospheric Pressure ....   Cloud Cover..... oktas/overcast   Temperature ....

Cloud Types ....

General Observations ....

Predictions for Tomorrow ....

Thursday

Date
and Time .... Location ....

Relative Humidity ....   Wind Speed and Direction ....

Atmospheric Pressure ....   Cloud Cover..... oktas/overcast   Temperature ....

Cloud Types ....

General Observations ....

Predictions for Tomorrow ....

Friday

Date
and Time .... Location ....

Relative Humidity ....   Wind Speed and Direction ....
Atmospheric Pressure ....   Cloud Cover..... oktas/overcast   Temperature ....

Cloud Types   ....

General Observations   ....

Predictions for Tomorrow   ....

Saturday

Date
and Time .... Location ....

Relative Humidity ....   Wind Speed and Direction ....
Atmospheric Pressure ....   Cloud Cover..... oktas/overcast   Temperature ....

Cloud Types   ....

General Observations   ....

Predictions for Tomorrow   ....

Sunday

Date
and Time .... Location ....

Relative Humidity ....   Wind Speed and Direction ....
Atmospheric Pressure ....   Cloud Cover..... oktas/overcast   Temperature ....

Cloud Types
....

General Observations
....

Predictions for Tomorrow
....

Weather Predictor's Tip

When you observe clouds from the ground, what you perceive is not necessarily
what is really happening in the sky. A cloud will appear to rise from the horizon
slowly; as it moves overhead it will seem to gradually increase in speed and size
until it reaches a maximum as it passes overhead and then begins to slow and
shrink as it appears to "sink" toward the opposite horizon.

Monday

Date
and Time .... Location ....

Relative Humidity .... Wind Speed and Direction ....
Atmospheric Pressure .... Cloud Cover..... oktas/overcast Temperature ....

Cloud Types ....

General Observations ....

Predictions for Tomorrow ....

Tuesday

Date
and Time .... Location ....

Relative Humidity .... Wind Speed and Direction ....
Atmospheric Pressure .... Cloud Cover..... oktas/overcast Temperature ....

Cloud Types ....

General Observations ....

Predictions for Tomorrow ....

Wednesday

Date
and Time .... Location ....

Relative Humidity .... Wind Speed and Direction ....
Atmospheric Pressure .... Cloud Cover..... oktas/overcast Temperature ....

Cloud Types ....

General Observations ....

Predictions for Tomorrow ....

Thursday

Date
and Time .... Location ....

Relative Humidity .... Wind Speed and Direction ....
Atmospheric Pressure .... Cloud Cover..... oktas/overcast Temperature ....

Cloud Types ....

General Observations ....

Predictions for Tomorrow ....

Friday

Date
and Time .... Location ....

Relative Humidity ....   Wind Speed and Direction ....

Atmospheric Pressure ....   Cloud Cover..... oktas/overcast   Temperature ....

Cloud Types   ....

General Observations   ....

Predictions for Tomorrow   ....

Saturday

Date
and Time .... Location ....

Relative Humidity ....   Wind Speed and Direction ....

Atmospheric Pressure ....   Cloud Cover..... oktas/overcast   Temperature ....

Cloud Types   ....

General Observations   ....

Predictions for Tomorrow   ....

Sunday

Date
and Time .... Location ....

Relative Humidity ....   Wind Speed and Direction ....

Atmospheric Pressure ....   Cloud Cover..... oktas/overcast   Temperature ....

Cloud Types   ....

General Observations   ....

Predictions for Tomorrow   ....

Weather Predictor's Tip

As rain moves through an area, any trees, buildings, or hills will become obscured by a shade of white. Over a body of water, the sheet of rain can be observed in the distance by looking for a change in color: that's where the precipitation meets the ocean. By tracking the movement of this boundary where the cascade meets the ground, lake, or ocean, you will be able to determine if and when the rain will strike your area.

Monday

Date and Time .... Location ....

Relative Humidity .... 　　Wind Speed and Direction ....

Atmospheric Pressure .... 　　Cloud Cover..... oktas/overcast 　Temperature ....

Cloud Types ....

General Observations ....

Predictions for Tomorrow ....

Tuesday

Date and Time .... Location ....

Relative Humidity .... 　　Wind Speed and Direction ....

Atmospheric Pressure .... 　　Cloud Cover..... oktas/overcast 　Temperature ....

Cloud Types ....

General Observations ....

Predictions for Tomorrow ....

Wednesday

Date and Time .... Location ....

Relative Humidity .... 　　Wind Speed and Direction ....

Atmospheric Pressure .... 　　Cloud Cover..... oktas/overcast 　Temperature ....

Cloud Types ....

General Observations ....

Predictions for Tomorrow ....

Thursday

Date and Time .... Location ....

Relative Humidity .... 　　Wind Speed and Direction ....

Atmospheric Pressure .... 　　Cloud Cover..... oktas/overcast 　Temperature ....

Cloud Types ....

General Observations ....

Predictions for Tomorrow ....

Friday

Date and Time .... Location ....

Relative Humidity ....   Wind Speed and Direction ....

Atmospheric Pressure ....   Cloud Cover..... oktas/overcast   Temperature ....

Cloud Types ....

General Observations ....

Predictions for Tomorrow ....

Saturday

Date and Time .... Location ....

Relative Humidity ....   Wind Speed and Direction ....

Atmospheric Pressure ....   Cloud Cover..... oktas/overcast   Temperature ....

Cloud Types ....

General Observations ....

Predictions for Tomorrow ....

Sunday

Date and Time .... Location ....

Relative Humidity ....   Wind Speed and Direction ....

Atmospheric Pressure ....   Cloud Cover..... oktas/overcast   Temperature ....

Cloud Types ....

General Observations ....

Predictions for Tomorrow ....

Weather Predictor's Tip

If the initial drops in a rainfall are medium to large, moderate to heavy rain may follow. If tiny drops of rain fall initially, drizzle or showers are probable. To determine the size of the drops of rain, you should observe how far they spread on surfaces such as cement. Another method is to observe the length of the drops when striking vertical surfaces such as windows.

Monday

Date and Time .... Location ....

Relative Humidity .... Wind Speed and Direction ....

Atmospheric Pressure .... Cloud Cover..... oktas/overcast Temperature ....

Cloud Types ....

General Observations ....

Predictions for Tomorrow ....

Tuesday

Date and Time .... Location ....

Relative Humidity .... Wind Speed and Direction ....

Atmospheric Pressure .... Cloud Cover..... oktas/overcast Temperature ....

Cloud Types ....

General Observations ....

Predictions for Tomorrow ....

Wednesday

Date and Time .... Location ....

Relative Humidity .... Wind Speed and Direction ....

Atmospheric Pressure .... Cloud Cover..... oktas/overcast Temperature ....

Cloud Types ....

General Observations ....

Predictions for Tomorrow ....

Thursday

Date and Time .... Location ....

Relative Humidity .... Wind Speed and Direction ....

Atmospheric Pressure .... Cloud Cover..... oktas/overcast Temperature ....

Cloud Types ....

General Observations ....

Predictions for Tomorrow ....

Friday

Date
and Time .... Location ....

Relative Humidity .... Wind Speed and Direction ....

Atmospheric Pressure .... Cloud Cover..... oktas/overcast Temperature ....

Cloud Types ....

General Observations ....

Predictions for Tomorrow ....

Saturday

Date
and Time .... Location ....

Relative Humidity .... Wind Speed and Direction ....

Atmospheric Pressure .... Cloud Cover..... oktas/overcast Temperature ....

Cloud Types ....

General Observations ....

Predictions for Tomorrow ....

Sunday

Date
and Time .... Location ....

Relative Humidity .... Wind Speed and Direction ....

Atmospheric Pressure .... Cloud Cover..... oktas/overcast Temperature ....

Cloud Types ....

General Observations ....

Predictions for Tomorrow ....

Weather Predictor's Tip

From the ground, the higher clouds—cirrus, cirrocumulus, and cirrostratus—seem to be more sharply defined than lower clouds. That is because the higher clouds contain more ice crystals, which gives them their distinctive clearly delineated aspect. High clouds are white except close to sunrise and sunset, when their bases reflect red, orange, and yellow wavelengths.

Monday

Date
and Time .... Location ....

Relative Humidity ....   Wind Speed and Direction ....

Atmospheric Pressure ....   Cloud Cover..... oktas/overcast   Temperature ....

Cloud Types ....

General Observations ....

Predictions for Tomorrow ....

Tuesday

Date
and Time .... Location ....

Relative Humidity ....   Wind Speed and Direction ....

Atmospheric Pressure ....   Cloud Cover..... oktas/overcast   Temperature ....

Cloud Types ....

General Observations ....

Predictions for Tomorrow ....

Wednesday

Date
and Time .... Location ....

Relative Humidity ....   Wind Speed and Direction ....

Atmospheric Pressure ....   Cloud Cover..... oktas/overcast   Temperature ....

Cloud Types ....

General Observations ....

Predictions for Tomorrow ....

Thursday

Date
and Time .... Location ....

Relative Humidity ....   Wind Speed and Direction ....

Atmospheric Pressure ....   Cloud Cover..... oktas/overcast   Temperature ....

Cloud Types ....

General Observations ....

Predictions for Tomorrow ....

Friday

Date and Time .... Location ....

Relative Humidity .... 　　Wind Speed and Direction ....

Atmospheric Pressure .... 　　Cloud Cover..... oktas/overcast　Temperature ....

Cloud Types ....

General Observations ....

Predictions for Tomorrow ....

Saturday

Date and Time .... Location ....

Relative Humidity .... 　　Wind Speed and Direction ....

Atmospheric Pressure .... 　　Cloud Cover..... oktas/overcast　Temperature ....

Cloud Types ....

General Observations ....

Predictions for Tomorrow ....

Sunday

Date and Time .... Location ....

Relative Humidity .... 　　Wind Speed and Direction ....

Atmospheric Pressure .... 　　Cloud Cover..... oktas/overcast　Temperature ....

Cloud Types ....

General Observations ....

Predictions for Tomorrow ....

Weather Predictor's Tip

The movement of fast-moving low clouds normally indicates the approximate direction of the wind. If you are near or on a lake or ocean, it is possible to observe surface flow of wind. Stronger winds will disturb the water surface, causing ripples. This region of ripples or darker texture can be observed moving as the wind progresses. Wind speed also creates different heights of waves.

Monday

Date and Time .... Location ....

Relative Humidity ....   Wind Speed and Direction ....

Atmospheric Pressure ....   Cloud Cover..... oktas/overcast   Temperature ....

Cloud Types ....

General Observations ....

Predictions for Tomorrow ....

Tuesday

Date and Time .... Location ....

Relative Humidity ....   Wind Speed and Direction ....

Atmospheric Pressure ....   Cloud Cover..... oktas/overcast   Temperature ....

Cloud Types ....

General Observations ....

Predictions for Tomorrow ....

Wednesday

Date and Time .... Location ....

Relative Humidity ....   Wind Speed and Direction ....

Atmospheric Pressure ....   Cloud Cover..... oktas/overcast   Temperature ....

Cloud Types ....

General Observations ....

Predictions for Tomorrow ....

Thursday

Date and Time .... Location ....

Relative Humidity ....   Wind Speed and Direction ....

Atmospheric Pressure ....   Cloud Cover..... oktas/overcast   Temperature ....

Cloud Types ....

General Observations ....

Predictions for Tomorrow ....

Friday

Date and Time .... Location ....

Relative Humidity ....   Wind Speed and Direction ....

Atmospheric Pressure ....   Cloud Cover..... oktas/overcast   Temperature ....

Cloud Types ....

General Observations ....

Predictions for Tomorrow ....

Saturday

Date and Time .... Location ....

Relative Humidity ....   Wind Speed and Direction ....

Atmospheric Pressure ....   Cloud Cover..... oktas/overcast   Temperature ....

Cloud Types ....

General Observations ....

Predictions for Tomorrow ....

Sunday

Date and Time .... Location ....

Relative Humidity ....   Wind Speed and Direction ....

Atmospheric Pressure ....   Cloud Cover..... oktas/overcast   Temperature ....

Cloud Types ....

General Observations ....

Predictions for Tomorrow ....

Weather Predictor's Tip

The frost line is the average depth to which the ground freezes during the winter in a given area. Frost lines are deeper in colder climates. Frost lines are of particular concern to contractors because building codes require that foundations be dug at least to the depth of the frost line to ensure structural integrity.

Monday

Date
and Time .... Location ....

Relative Humidity .... Wind Speed and Direction ....
Atmospheric Pressure .... Cloud Cover..... oktas/overcast Temperature ....

Cloud Types ....

General Observations ....

Predictions for Tomorrow ....

Tuesday

Date
and Time .... Location ....

Relative Humidity .... Wind Speed and Direction ....
Atmospheric Pressure .... Cloud Cover..... oktas/overcast Temperature ....

Cloud Types ....

General Observations ....

Predictions for Tomorrow ....

Wednesday

Date
and Time .... Location ....

Relative Humidity .... Wind Speed and Direction ....
Atmospheric Pressure .... Cloud Cover..... oktas/overcast Temperature ....

Cloud Types ....

General Observations ....

Predictions for Tomorrow ....

Thursday

Date
and Time .... Location ....

Relative Humidity .... Wind Speed and Direction ....
Atmospheric Pressure .... Cloud Cover..... oktas/overcast Temperature ....

Cloud Types ....

General Observations ....

Predictions for Tomorrow ....

Friday

Date and Time .... Location ....

Relative Humidity ....   Wind Speed and Direction ....

Atmospheric Pressure ....   Cloud Cover..... oktas/overcast   Temperature ....

Cloud Types ....

General Observations ....

Predictions for Tomorrow ....

Saturday

Date and Time .... Location ....

Relative Humidity ....   Wind Speed and Direction ....

Atmospheric Pressure ....   Cloud Cover..... oktas/overcast   Temperature ....

Cloud Types ....

General Observations ....

Predictions for Tomorrow ....

Sunday

Date and Time .... Location ....

Relative Humidity ....   Wind Speed and Direction ....

Atmospheric Pressure ....   Cloud Cover..... oktas/overcast   Temperature ....

Cloud Types ....

General Observations ....

Predictions for Tomorrow ....

Weather Predictor's Tip

Water runoff along the sides of roads indicates moderate to heavy rain within the past hour or so. Large puddles along the side of the road are an indication of earlier moderate to heavy rain. Long periods of moderate showers can also create puddles. An observer must take into account the slope or curvature of the road. The type of surface also affects the rate of evaporation.

Monday

Date
and Time .... Location ....

Relative Humidity .... Wind Speed and Direction ....
Atmospheric Pressure .... Cloud Cover..... oktas/overcast   Temperature ....

Cloud Types ....

General Observations ....

Predictions for Tomorrow ....

Tuesday

Date
and Time .... Location ....

Relative Humidity .... Wind Speed and Direction ....
Atmospheric Pressure .... Cloud Cover..... oktas/overcast   Temperature ....

Cloud Types ....

General Observations ....

Predictions for Tomorrow ....

Wednesday

Date
and Time .... Location ....

Relative Humidity .... Wind Speed and Direction ....
Atmospheric Pressure .... Cloud Cover..... oktas/overcast   Temperature ....

Cloud Types ....

General Observations ....

Predictions for Tomorrow ....

Thursday

Date
and Time .... Location ....

Relative Humidity .... Wind Speed and Direction ....
Atmospheric Pressure .... Cloud Cover..... oktas/overcast   Temperature ....

Cloud Types ....

General Observations ....

Predictions for Tomorrow ....

Friday

Date
and Time .... Location ....

Relative Humidity ....   Wind Speed and Direction ....

Atmospheric Pressure ....   Cloud Cover..... oktas/overcast   Temperature ....

Cloud Types ....

General Observations ....

Predictions for Tomorrow ....

Saturday

Date
and Time .... Location ....

Relative Humidity ....   Wind Speed and Direction ....

Atmospheric Pressure ....   Cloud Cover..... oktas/overcast   Temperature ....

Cloud Types ....

General Observations ....

Predictions for Tomorrow ....

Sunday

Date
and Time .... Location ....

Relative Humidity ....   Wind Speed and Direction ....

Atmospheric Pressure ....   Cloud Cover..... oktas/overcast   Temperature ....

Cloud Types ....

General Observations ....

Predictions for Tomorrow ....

Weather Predictor's Tip

A thunderstorm is classified as severe by the United States National Weather Service if it meets at least one of three criteria: wind speeds reach 58 mph (92 km/h) or above, hail is three quarters of an inch (2 cm) in diameter or larger, or the thunderstorm produces at least one tornado. In the United States, flash floods from thunderstorms cause more deaths than tornadoes.

Monday

Date and Time .... Location ....

Relative Humidity ....   Wind Speed and Direction ....

Atmospheric Pressure ....   Cloud Cover..... oktas/overcast   Temperature ....

Cloud Types ....

General Observations ....

Predictions for Tomorrow ....

Tuesday

Date and Time .... Location ....

Relative Humidity ....   Wind Speed and Direction ....

Atmospheric Pressure ....   Cloud Cover..... oktas/overcast   Temperature ....

Cloud Types ....

General Observations ....

Predictions for Tomorrow ....

Wednesday

Date and Time .... Location ....

Relative Humidity ....   Wind Speed and Direction ....

Atmospheric Pressure ....   Cloud Cover..... oktas/overcast   Temperature ....

Cloud Types ....

General Observations ....

Predictions for Tomorrow ....

Thursday

Date and Time .... Location ....

Relative Humidity ....   Wind Speed and Direction ....

Atmospheric Pressure ....   Cloud Cover..... oktas/overcast   Temperature ....

Cloud Types ....

General Observations ....

Predictions for Tomorrow ....

880

89

Friday

Date
and Time .... Location ....

Relative Humidity ....   Wind Speed and Direction ....

Atmospheric Pressure ....   Cloud Cover..... oktas/overcast   Temperature ....

Cloud Types ....

General Observations ....

Predictions for Tomorrow ....

Saturday

Date
and Time .... Location ....

Relative Humidity ....   Wind Speed and Direction ....

Atmospheric Pressure ....   Cloud Cover..... oktas/overcast   Temperature ....

Cloud Types ....

General Observations ....

Predictions for Tomorrow ....

Sunday

Date
and Time .... Location ....

Relative Humidity ....   Wind Speed and Direction ....

Atmospheric Pressure ....   Cloud Cover..... oktas/overcast   Temperature ....

Cloud Types ....

General Observations ....

Predictions for Tomorrow ....

Weather Predictor's Tip

The strength of winds can be observed even after they have passed. Plants—long grass especially—are often flattened. If the flattened plants are surrounded by a field that is otherwise undisturbed, then the wind was most likely from a downdraft produced by a thunderstorm. Trees stripped of their leaves and twigs are another indicator of strong winds, particularly if large branches have snapped.

Monday

Date and Time .... Location ....

Relative Humidity ....   Wind Speed and Direction ....

Atmospheric Pressure ....   Cloud Cover..... oktas/overcast   Temperature ....

Cloud Types ....

General Observations ....

Predictions for Tomorrow ....

Tuesday

Date and Time .... Location ....

Relative Humidity ....   Wind Speed and Direction ....

Atmospheric Pressure ....   Cloud Cover..... oktas/overcast   Temperature ....

Cloud Types ....

General Observations ....

Predictions for Tomorrow ....

Wednesday

Date and Time .... Location ....

Relative Humidity ....   Wind Speed and Direction ....

Atmospheric Pressure ....   Cloud Cover..... oktas/overcast   Temperature ....

Cloud Types ....

General Observations ....

Predictions for Tomorrow ....

Thursday

Date and Time .... Location ....

Relative Humidity ....   Wind Speed and Direction ....

Atmospheric Pressure ....   Cloud Cover..... oktas/overcast   Temperature ....

Cloud Types ....

General Observations ....

Predictions for Tomorrow ....

Friday

Date
and Time .... Location ....

Relative Humidity .... Wind Speed and Direction ....
Atmospheric Pressure .... Cloud Cover..... oktas/overcast Temperature ....

Cloud Types ....

General Observations ....

Predictions for Tomorrow ....

Saturday

Date
and Time .... Location ....

Relative Humidity .... Wind Speed and Direction ....
Atmospheric Pressure .... Cloud Cover..... oktas/overcast Temperature ....

Cloud Types ....

General Observations ....

Predictions for Tomorrow ....

Sunday

Date
and Time .... Location ....

Relative Humidity .... Wind Speed and Direction ....
Atmospheric Pressure .... Cloud Cover..... oktas/overcast Temperature ....

Cloud Types ....

General Observations ....

Predictions for Tomorrow ....

Weather Predictor's Tip

From the front of a thunderstorm, you can see the thickening cirrostratus and altostratus clouds as their anvil top approaches. The storm will gradually become darker; then rain will appear and lightning become more frequent. If one side of the anvil is spreading more than the other it means that the storm is approaching at an angle and will veer in the direction of the extended side.

Monday

Date and Time .... Location ....

Relative Humidity ....   Wind Speed and Direction ....

Atmospheric Pressure ....   Cloud Cover..... oktas/overcast   Temperature ....

Cloud Types ....

General Observations ....

Predictions for Tomorrow ....

Tuesday

Date and Time .... Location ....

Relative Humidity ....   Wind Speed and Direction ....

Atmospheric Pressure ....   Cloud Cover..... oktas/overcast   Temperature ....

Cloud Types ....

General Observations ....

Predictions for Tomorrow ....

Wednesday

Date and Time .... Location ....

Relative Humidity ....   Wind Speed and Direction ....

Atmospheric Pressure ....   Cloud Cover..... oktas/overcast   Temperature ....

Cloud Types ....

General Observations ....

Predictions for Tomorrow ....

Thursday

Date and Time .... Location ....

Relative Humidity ....   Wind Speed and Direction ....

Atmospheric Pressure ....   Cloud Cover..... oktas/overcast   Temperature ....

Cloud Types ....

General Observations ....

Predictions for Tomorrow ....

Friday

Date
and Time .... Location ....

Relative Humidity ....   Wind Speed and Direction ....

Atmospheric Pressure ....   Cloud Cover..... oktas/overcast   Temperature ....

Cloud Types ....

General Observations ....

Predictions for Tomorrow ....

Saturday

Date
and Time .... Location ....

Relative Humidity ....   Wind Speed and Direction ....

Atmospheric Pressure ....   Cloud Cover..... oktas/overcast   Temperature ....

Cloud Types ....

General Observations ....

Predictions for Tomorrow ....

Sunday

Date
and Time .... Location ....

Relative Humidity ....   Wind Speed and Direction ....

Atmospheric Pressure ....   Cloud Cover..... oktas/overcast   Temperature ....

Cloud Types ....

General Observations ....

Predictions for Tomorrow ....

Weather Predictor's Tip

Visual opacity is a technical term to describe how much sunlight is getting through a cloud. Clouds are considered opaque if they are so thick that they blot out the Sun. Clouds are translucent when the Sun can be perceived dimly through the cloud as a silhouette. Clouds are transparent if the Sun can easily be seen through the cloud. Under transparent clouds, objects will typically cast shadows.

Monday

Date and Time .... Location ....

Relative Humidity ....   Wind Speed and Direction ....

Atmospheric Pressure ....   Cloud Cover..... oktas/overcast   Temperature ....

Cloud Types ....

General Observations ....

Predictions for Tomorrow ....

Tuesday

Date and Time .... Location ....

Relative Humidity ....   Wind Speed and Direction ....

Atmospheric Pressure ....   Cloud Cover..... oktas/overcast   Temperature ....

Cloud Types ....

General Observations ....

Predictions for Tomorrow ....

Wednesday

Date and Time .... Location ....

Relative Humidity ....   Wind Speed and Direction ....

Atmospheric Pressure ....   Cloud Cover..... oktas/overcast   Temperature ....

Cloud Types ....

General Observations ....

Predictions for Tomorrow ....

Thursday

Date and Time .... Location ....

Relative Humidity ....   Wind Speed and Direction ....

Atmospheric Pressure ....   Cloud Cover..... oktas/overcast   Temperature ....

Cloud Types ....

General Observations ....

Predictions for Tomorrow ....

1029   1027

Friday

Date
and Time .... Location ....

Relative Humidity ....   Wind Speed and Direction ....
Atmospheric Pressure ....   Cloud Cover..... oktas/overcast   Temperature ....

Cloud Types   ....

General Observations ....

Predictions for Tomorrow   ....

Saturday

Date
and Time .... Location ....

Relative Humidity ....   Wind Speed and Direction ....
Atmospheric Pressure ....   Cloud Cover..... oktas/overcast   Temperature ....

Cloud Types   ....

General Observations ....

Predictions for Tomorrow   ....

Sunday

Date
and Time .... Location ....

Relative Humidity ....   Wind Speed and Direction ....
Atmospheric Pressure ....   Cloud Cover..... oktas/overcast   Temperature ....

Cloud Types   ....

General Observations ....

Predictions for Tomorrow   ....

Weather Predictor's Tip

When atmospheric conditions are unstable, many cumulus clouds will develop but only a few will dominate. That is because the updrafts of a few dominant clouds are sufficiently strong to remove air from the surroundings, depriving other nearby clouds of their updrafts to further develop. Winds at the top of the clouds are usually stronger, which can cause the towering cumulus to lean forward slightly as it develops.

Monday

Date and Time .... Location ....

Relative Humidity ....   Wind Speed and Direction ....

Atmospheric Pressure ....   Cloud Cover..... oktas/overcast   Temperature ....

Cloud Types ....

General Observations ....

Predictions for Tomorrow ....

Tuesday

Date and Time .... Location ....

Relative Humidity ....   Wind Speed and Direction ....

Atmospheric Pressure ....   Cloud Cover..... oktas/overcast   Temperature ....

Cloud Types ....

General Observations ....

Predictions for Tomorrow ....

Wednesday

Date and Time .... Location ....

Relative Humidity ....   Wind Speed and Direction ....

Atmospheric Pressure ....   Cloud Cover..... oktas/overcast   Temperature ....

Cloud Types ....

General Observations ....

Predictions for Tomorrow ....

Thursday

Date and Time .... Location ....

Relative Humidity ....   Wind Speed and Direction ....

Atmospheric Pressure ....   Cloud Cover..... oktas/overcast   Temperature ....

Cloud Types ....

General Observations ....

Predictions for Tomorrow ....

Friday

Date and Time .... Location ....

Relative Humidity .... Wind Speed and Direction ....

Atmospheric Pressure .... Cloud Cover..... oktas/overcast   Temperature ....

Cloud Types   ....

General Observations   ....

Predictions for Tomorrow   ....

Saturday

Date and Time .... Location ....

Relative Humidity .... Wind Speed and Direction ....

Atmospheric Pressure .... Cloud Cover..... oktas/overcast   Temperature ....

Cloud Types   ....

General Observations   ....

Predictions for Tomorrow   ....

Sunday

Date and Time .... Location ....

Relative Humidity .... Wind Speed and Direction ....

Atmospheric Pressure .... Cloud Cover..... oktas/overcast   Temperature ....

Cloud Types   ....

General Observations   ....

Predictions for Tomorrow   ....

Weather Predictor's Tip

Water can exist in three forms—as water vapor, as liquid, and as ice—but always consists of two hydrogen atoms and one oxygen atom. However, there is a difference in the energy of each water molecule in each form: water vapor molecules exhibit the most motion, ice molecules display the least motion, while liquid water molecules have moderate energy.

Monday

Date
and Time .... Location ....

Relative Humidity ....   Wind Speed and Direction ....

Atmospheric Pressure ....   Cloud Cover..... oktas/overcast   Temperature ....

Cloud Types   ....

General Observations   ....

Predictions for Tomorrow   ....

Tuesday

Date
and Time .... Location ....

Relative Humidity ....   Wind Speed and Direction ....

Atmospheric Pressure ....   Cloud Cover..... oktas/overcast   Temperature ....

Cloud Types   ....

General Observations   ....

Predictions for Tomorrow   ....

Wednesday

Date
and Time .... Location ....

Relative Humidity ....   Wind Speed and Direction ....

Atmospheric Pressure ....   Cloud Cover..... oktas/overcast   Temperature ....

Cloud Types   ....

General Observations   ....

Predictions for Tomorrow   ....

Thursday

Date
and Time .... Location ....

Relative Humidity ....   Wind Speed and Direction ....

Atmospheric Pressure ....   Cloud Cover..... oktas/overcast   Temperature ....

Cloud Types   ....

General Observations   ....

Predictions for Tomorrow   ....

Friday

Date
and Time .... Location ....

Relative Humidity ....   Wind Speed and Direction ....

Atmospheric Pressure ....   Cloud Cover..... oktas/overcast   Temperature ....

Cloud Types   ....

General Observations   ....

Predictions for Tomorrow   ....

Saturday

Date
and Time .... Location ....

Relative Humidity ....   Wind Speed and Direction ....

Atmospheric Pressure ....   Cloud Cover..... oktas/overcast   Temperature ....

Cloud Types   ....

General Observations   ....

Predictions for Tomorrow   ....

Sunday

Date
and Time .... Location ....

Relative Humidity ....   Wind Speed and Direction ....

Atmospheric Pressure ....   Cloud Cover..... oktas/overcast   Temperature ....

Cloud Types   ....

General Observations   ....

Predictions for Tomorrow   ....

Weather Predictor's Tip

A red moon does not mean that a change in weather is coming. All it means is that there is a high concentration of particles in the air such as dust and smoke that act to scatter short and intermediate wavelengths of light (violet, blue, and yellow). As a result, we can only see the longer wavelengths of orange and red.

Monday

Date and Time .... Location ....

Relative Humidity ....   Wind Speed and Direction ....

Atmospheric Pressure ....   Cloud Cover..... oktas/overcast   Temperature ....

Cloud Types   ....

General Observations   ....

Predictions for Tomorrow   ....

Tuesday

Date and Time .... Location ....

Relative Humidity ....   Wind Speed and Direction ....

Atmospheric Pressure ....   Cloud Cover..... oktas/overcast   Temperature ....

Cloud Types   ....

General Observations   ....

Predictions for Tomorrow   ....

Wednesday

Date and Time .... Location ....

Relative Humidity ....   Wind Speed and Direction ....

Atmospheric Pressure ....   Cloud Cover..... oktas/overcast   Temperature ....

Cloud Types   ....

General Observations   ....

Predictions for Tomorrow   ....

Thursday

Date and Time .... Location ....

Relative Humidity ....   Wind Speed and Direction ....

Atmospheric Pressure ....   Cloud Cover..... oktas/overcast   Temperature ....

Cloud Types   ....

General Observations   ....

Predictions for Tomorrow   ....

Friday

Date
and Time .... Location ....

Relative Humidity ....   Wind Speed and Direction ....

Atmospheric Pressure ....   Cloud Cover..... oktas/overcast   Temperature ....

Cloud Types ....

General Observations ....

Predictions for Tomorrow ....

Saturday

Date
and Time .... Location ....

Relative Humidity ....   Wind Speed and Direction ....

Atmospheric Pressure ....   Cloud Cover..... oktas/overcast   Temperature ....

Cloud Types ....

General Observations ....

Predictions for Tomorrow ....

Sunday

Date
and Time .... Location ....

Relative Humidity ....   Wind Speed and Direction ....

Atmospheric Pressure ....   Cloud Cover..... oktas/overcast   Temperature ....

Cloud Types ....

General Observations ....

Predictions for Tomorrow ....

Weather Predictor's Tip

Calculating the amount of rain to snow will depend on prevailing conditions. If the air is warm, it may take the equivalent of as much as 5 inches (12.5 cm) of rain to get 10 inches (25 cm) of wet snow because the snow may pack or melt quickly. When the air is very cold, the ratio may fall to about 1/2 inch (1.5 cm) of rain to 10 inches (25 cm) of "dry" fluffy snow.

Monday

Date and Time .... Location ....

Relative Humidity ....   Wind Speed and Direction ....

Atmospheric Pressure ....   Cloud Cover..... oktas/overcast   Temperature ....

Cloud Types   ....

General Observations   ....

Predictions for Tomorrow   ....

Tuesday

Date and Time .... Location ....

Relative Humidity ....   Wind Speed and Direction ....

Atmospheric Pressure ....   Cloud Cover..... oktas/overcast   Temperature ....

Cloud Types   ....

General Observations   ....

Predictions for Tomorrow   ....

Wednesday

Date and Time .... Location ....

Relative Humidity ....   Wind Speed and Direction ....

Atmospheric Pressure ....   Cloud Cover..... oktas/overcast   Temperature ....

Cloud Types   ....

General Observations   ....

Predictions for Tomorrow   ....

Thursday

Date and Time .... Location ....

Relative Humidity ....   Wind Speed and Direction ....

Atmospheric Pressure ....   Cloud Cover..... oktas/overcast   Temperature ....

Cloud Types   ....

General Observations   ....

Predictions for Tomorrow   ....

Friday

Date
and Time .... Location ....

Relative Humidity .... Wind Speed and Direction ....
Atmospheric Pressure .... Cloud Cover..... oktas/overcast Temperature ....

Cloud Types ....

General Observations ....

Predictions for Tomorrow ....

Saturday

Date
and Time .... Location ....

Relative Humidity .... Wind Speed and Direction ....
Atmospheric Pressure .... Cloud Cover..... oktas/overcast Temperature ....

Cloud Types ....

General Observations ....

Predictions for Tomorrow ....

Sunday

Date
and Time .... Location ....

Relative Humidity .... Wind Speed and Direction ....
Atmospheric Pressure .... Cloud Cover..... oktas/overcast Temperature ....

Cloud Types ....

General Observations ....

Predictions for Tomorrow ....

Weather Predictor's Tip

The average raindrop is about 0.04 to 0.08 of an inch (1–2 mm) in diameter.
The larger a raindrop, the faster it will fall. A raindrop with a diameter of
0.04 in (1 mm) will typically reach a speed of about 9 mph (14 km/h) as it
falls to Earth, but a drop of 0.2 in (5 mm) can fall at speeds of 20 mph
(32 km/h).

Monday

Date
and Time .... Location ....

Relative Humidity .... Wind Speed and Direction ....

Atmospheric Pressure .... Cloud Cover..... oktas/overcast   Temperature ....

Cloud Types ....

General Observations ....

Predictions for Tomorrow ....

Tuesday

Date
and Time .... Location ....

Relative Humidity .... Wind Speed and Direction ....

Atmospheric Pressure .... Cloud Cover..... oktas/overcast   Temperature ....

Cloud Types ....

General Observations ....

Predictions for Tomorrow ....

Wednesday

Date
and Time .... Location ....

Relative Humidity .... Wind Speed and Direction ....

Atmospheric Pressure .... Cloud Cover..... oktas/overcast   Temperature ....

Cloud Types ....

General Observations ....

Predictions for Tomorrow ....

Thursday

Date
and Time .... Location ....

Relative Humidity .... Wind Speed and Direction ....

Atmospheric Pressure .... Cloud Cover..... oktas/overcast   Temperature ....

Cloud Types ....

General Observations ....

Predictions for Tomorrow ....

Friday

Date and Time .... Location ....

Relative Humidity ....   Wind Speed and Direction ....

Atmospheric Pressure ....   Cloud Cover..... oktas/overcast   Temperature ....

Cloud Types ....

General Observations ....

Predictions for Tomorrow ....

Saturday

Date and Time .... Location ....

Relative Humidity ....   Wind Speed and Direction ....

Atmospheric Pressure ....   Cloud Cover..... oktas/overcast   Temperature ....

Cloud Types ....

General Observations ....

Predictions for Tomorrow ....

Sunday

Date and Time .... Location ....

Relative Humidity ....   Wind Speed and Direction ....

Atmospheric Pressure ....   Cloud Cover..... oktas/overcast   Temperature ....

Cloud Types ....

General Observations ....

Predictions for Tomorrow ....

Weather Predictor's Tip

Ice storms are caused by freezing rain. These dangerous storms, which can cause traffic accidents and snap power lines, contain very strong updrafts, reaching up to 100 mph (160 km/h). These updrafts can suspend rain in mid-air in temperatures below freezing, causing rain to turn into hail. One of the worst ice storms occurred in Canada over six days in January 1998, leaving accumulations of 3–4 inches (7–11 cm) of ice.

Monday

Date and Time .... Location ....

Relative Humidity .... Wind Speed and Direction ....

Atmospheric Pressure .... Cloud Cover..... oktas/overcast Temperature ....

Cloud Types ....

General Observations ....

Predictions for Tomorrow ....

Tuesday

Date and Time .... Location ....

Relative Humidity .... Wind Speed and Direction ....

Atmospheric Pressure .... Cloud Cover..... oktas/overcast Temperature ....

Cloud Types ....

General Observations ....

Predictions for Tomorrow ....

Wednesday

Date and Time .... Location ....

Relative Humidity .... Wind Speed and Direction ....

Atmospheric Pressure .... Cloud Cover..... oktas/overcast Temperature ....

Cloud Types ....

General Observations ....

Predictions for Tomorrow ....

Thursday

Date and Time .... Location ....

Relative Humidity .... Wind Speed and Direction ....

Atmospheric Pressure .... Cloud Cover..... oktas/overcast Temperature ....

Cloud Types ....

General Observations ....

Predictions for Tomorrow ....

880

89

Friday

Date and Time .... Location ....

Relative Humidity .... Wind Speed and Direction ....

Atmospheric Pressure .... Cloud Cover..... oktas/overcast Temperature ....

Cloud Types ....

General Observations ....

Predictions for Tomorrow ....

Saturday

Date and Time .... Location ....

Relative Humidity .... Wind Speed and Direction ....

Atmospheric Pressure .... Cloud Cover..... oktas/overcast Temperature ....

Cloud Types ....

General Observations ....

Predictions for Tomorrow ....

Sunday

Date and Time .... Location ....

Relative Humidity .... Wind Speed and Direction ....

Atmospheric Pressure .... Cloud Cover..... oktas/overcast Temperature ....

Cloud Types ....

General Observations ....

Predictions for Tomorrow ....

Weather Predictor's Tip

Large bodies of water such as oceans and lakes offer a major source of moisture for clouds, fog, and precipitation. When cold air blows over a lake, for instance, moisture from the warmer lake water rises, cools, and condenses into clouds. This is called the "lake effect." The lake effect can also occur over salt water.

880

89(

Monday

Date
and Time .... Location ....

Relative Humidity ....   Wind Speed and Direction ....

Atmospheric Pressure ....   Cloud Cover..... oktas/overcast   Temperature ....

Cloud Types   ....

General Observations   ....

Predictions for Tomorrow   ....

Tuesday

Date
and Time .... Location ....

Relative Humidity ....   Wind Speed and Direction ....

Atmospheric Pressure ....   Cloud Cover..... oktas/overcast   Temperature ....

Cloud Types   ....

General Observations   ....

Predictions for Tomorrow   ....

Wednesday

Date
and Time .... Location ....

Relative Humidity ....   Wind Speed and Direction ....

Atmospheric Pressure ....   Cloud Cover..... oktas/overcast   Temperature ....

Cloud Types   ....

General Observations   ....

Predictions for Tomorrow   ....

Thursday

Date
and Time .... Location ....

Relative Humidity ....   Wind Speed and Direction ....

Atmospheric Pressure ....   Cloud Cover..... oktas/overcast   Temperature ....

Cloud Types   ....

General Observations   ....

Predictions for Tomorrow   ....

Friday

Date
and Time .... Location ....

Relative Humidity ....   Wind Speed and Direction ....
Atmospheric Pressure ....   Cloud Cover..... oktas/overcast   Temperature ....

Cloud Types   ....

General Observations   ....

Predictions for Tomorrow   ....

Saturday

Date
and Time .... Location ....

Relative Humidity ....   Wind Speed and Direction ....
Atmospheric Pressure ....   Cloud Cover..... oktas/overcast   Temperature ....

Cloud Types   ....

General Observations   ....

Predictions for Tomorrow   ....

Sunday

Date
and Time .... Location ....

Relative Humidity ....   Wind Speed and Direction ....
Atmospheric Pressure ....   Cloud Cover..... oktas/overcast   Temperature ....

Cloud Types   ....

General Observations   ....

Predictions for Tomorrow   ....

Weather Predictor's Tip

A trough is indicated on a weather map by an elongated region of low atmospheric
pressure bordered by a broken line. Low-pressure regions are associated with
cloudy or stormy weather near and to the east of the trough. Higher-pressure
regions are found west of the trough, meaning air is sinking and fewer clouds
are forming.

Monday

Date and Time .... Location ....

Relative Humidity ....   Wind Speed and Direction ....

Atmospheric Pressure ....   Cloud Cover..... oktas/overcast   Temperature ....

Cloud Types ....

General Observations ....

Predictions for Tomorrow ....

Tuesday

Date and Time .... Location ....

Relative Humidity ....   Wind Speed and Direction ....

Atmospheric Pressure ....   Cloud Cover..... oktas/overcast   Temperature ....

Cloud Types ....

General Observations ....

Predictions for Tomorrow ....

Wednesday

Date and Time .... Location ....

Relative Humidity ....   Wind Speed and Direction ....

Atmospheric Pressure ....   Cloud Cover..... oktas/overcast   Temperature ....

Cloud Types ....

General Observations ....

Predictions for Tomorrow ....

Thursday

Date and Time .... Location ....

Relative Humidity ....   Wind Speed and Direction ....

Atmospheric Pressure ....   Cloud Cover..... oktas/overcast   Temperature ....

Cloud Types ....

General Observations ....

Predictions for Tomorrow ....

Friday

Date
and Time .... Location ....

Relative Humidity ....   Wind Speed and Direction ....

Atmospheric Pressure ....   Cloud Cover..... oktas/overcast   Temperature ....

Cloud Types ....

General Observations ....

Predictions for Tomorrow ....

Saturday

Date
and Time .... Location ....

Relative Humidity ....   Wind Speed and Direction ....

Atmospheric Pressure ....   Cloud Cover..... oktas/overcast   Temperature ....

Cloud Types ....

General Observations ....

Predictions for Tomorrow ....

Sunday

Date
and Time .... Location ....

Relative Humidity ....   Wind Speed and Direction ....

Atmospheric Pressure ....   Cloud Cover..... oktas/overcast   Temperature ....

Cloud Types ....

General Observations ....

Predictions for Tomorrow ....

Weather Predictor's Tip

"Partly sunny" and "partly cloudy" in a weather forecast both mean that 30 to 60 percent of the sky is covered by clouds. "Cloudy" means that 90 to 100 percent of the sky is covered; "mostly cloudy" that 70 to 80 percent is covered; "mostly sunny" or "mostly clear" that only 10 to 30 percent is covered; and "sunny" or "clear" mean that no more than 10 percent is covered.

Monday

Date
and Time .... Location ....

Relative Humidity ....   Wind Speed and Direction ....

Atmospheric Pressure ....   Cloud Cover..... oktas/overcast   Temperature ....

Cloud Types   ....

General Observations   ....

Predictions for Tomorrow   ....

Tuesday

Date
and Time .... Location ....

Relative Humidity ....   Wind Speed and Direction ....

Atmospheric Pressure ....   Cloud Cover..... oktas/overcast   Temperature ....

Cloud Types   ....

General Observations   ....

Predictions for Tomorrow   ....

Wednesday

Date
and Time .... Location ....

Relative Humidity ....   Wind Speed and Direction ....

Atmospheric Pressure ....   Cloud Cover..... oktas/overcast   Temperature ....

Cloud Types   ....

General Observations   ....

Predictions for Tomorrow   ....

Thursday

Date
and Time .... Location ....

Relative Humidity ....   Wind Speed and Direction ....

Atmospheric Pressure ....   Cloud Cover..... oktas/overcast   Temperature ....

Cloud Types   ....

General Observations   ....

Predictions for Tomorrow   ....

1029   1027

880

890

1029

Friday

Date
and Time .... Location ....

Relative Humidity ....   Wind Speed and Direction ....

Atmospheric Pressure ....   Cloud Cover..... oktas/overcast   Temperature ....

Cloud Types ....

General Observations ....

Predictions for Tomorrow ....

Saturday

Date
and Time .... Location ....

Relative Humidity ....   Wind Speed and Direction ....

Atmospheric Pressure ....   Cloud Cover..... oktas/overcast   Temperature ....

Cloud Types ....

General Observations ....

Predictions for Tomorrow ....

Sunday

Date
and Time .... Location ....

Relative Humidity ....   Wind Speed and Direction ....

Atmospheric Pressure ....   Cloud Cover..... oktas/overcast   Temperature ....

Cloud Types ....

General Observations ....

Predictions for Tomorrow ....

Weather Predictor's Tip

It is almost always warmer when the Sun is shining. That is why temperatures tend to be coldest at dawn after hours of darkness. But wind can cancel out the effect of sunlight by bringing cold air into a region, a process known as advection. Wind can also bring warmer air on a cloudy day. The solar influence on wind direction is limited to regions where the land meets the sea, producing sea and land breezes.

Monday

Date
and Time .... Location ....

Relative Humidity ....   Wind Speed and Direction ....

Atmospheric Pressure ....   Cloud Cover..... oktas/overcast   Temperature ....

Cloud Types   ....

General Observations   ....

Predictions for Tomorrow   ....

880

890

Tuesday

Date
and Time .... Location ....

Relative Humidity ....   Wind Speed and Direction ....

Atmospheric Pressure ....   Cloud Cover..... oktas/overcast   Temperature ....

Cloud Types   ....

General Observations   ....

Predictions for Tomorrow   ....

Wednesday

Date
and Time .... Location ....

Relative Humidity ....   Wind Speed and Direction ....

Atmospheric Pressure ....   Cloud Cover..... oktas/overcast   Temperature ....

Cloud Types   ....

General Observations   ....

Predictions for Tomorrow   ....

Thursday

Date
and Time .... Location ....

Relative Humidity ....   Wind Speed and Direction ....

Atmospheric Pressure ....   Cloud Cover..... oktas/overcast   Temperature ....

Cloud Types   ....

General Observations   ....

Predictions for Tomorrow   ....

Friday

Date and Time .... Location ....

Relative Humidity .... Wind Speed and Direction ....

Atmospheric Pressure .... Cloud Cover..... oktas/overcast   Temperature ....

Cloud Types ....

General Observations ....

Predictions for Tomorrow ....

Saturday

Date and Time .... Location ....

Relative Humidity .... Wind Speed and Direction ....

Atmospheric Pressure .... Cloud Cover..... oktas/overcast   Temperature ....

Cloud Types ....

General Observations ....

Predictions for Tomorrow ....

Sunday

Date and Time .... Location ....

Relative Humidity .... Wind Speed and Direction ....

Atmospheric Pressure .... Cloud Cover..... oktas/overcast   Temperature ....

Cloud Types ....

General Observations ....

Predictions for Tomorrow ....

Weather Predictor's Tip

The color of a cloud depends on three things: the Sun, the thickness of the cloud, and the position of the observer. A cloud filled with a lot of moisture will appear darker if the cloud is between the Sun and the person on the ground. A thick cloud will appear white if the Sun is to the observer's back. Mid- and low-level clouds show the most intense coloring 5 to 10 minutes after sunset or before sunrise.

Monday

Date and Time .... Location ....

Relative Humidity .... Wind Speed and Direction ....
Atmospheric Pressure .... Cloud Cover..... oktas/overcast Temperature ....

Cloud Types ....

General Observations ....

Predictions for Tomorrow ....

Tuesday

Date and Time .... Location ....

Relative Humidity .... Wind Speed and Direction ....
Atmospheric Pressure .... Cloud Cover..... oktas/overcast Temperature ....

Cloud Types ....

General Observations ....

Predictions for Tomorrow ....

Wednesday

Date and Time .... Location ....

Relative Humidity .... Wind Speed and Direction ....
Atmospheric Pressure .... Cloud Cover..... oktas/overcast Temperature ....

Cloud Types ....

General Observations ....

Predictions for Tomorrow ....

Thursday

Date and Time .... Location ....

Relative Humidity .... Wind Speed and Direction ....
Atmospheric Pressure .... Cloud Cover..... oktas/overcast Temperature ....

Cloud Types ....

General Observations ....

Predictions for Tomorrow ....

Friday

Date and Time .... Location ....

Relative Humidity ....   Wind Speed and Direction ....

Atmospheric Pressure ....   Cloud Cover..... oktas/overcast   Temperature ....

Cloud Types ....

General Observations ....

Predictions for Tomorrow ....

Saturday

Date and Time .... Location ....

Relative Humidity ....   Wind Speed and Direction ....

Atmospheric Pressure ....   Cloud Cover..... oktas/overcast   Temperature ....

Cloud Types ....

General Observations ....

Predictions for Tomorrow ....

Sunday

Date and Time .... Location ....

Relative Humidity ....   Wind Speed and Direction ....

Atmospheric Pressure ....   Cloud Cover..... oktas/overcast   Temperature ....

Cloud Types ....

General Observations ....

Predictions for Tomorrow ....

Weather Predictor's Tip

Virga consist of streaks or wisps of precipitation that fall from clouds but evaporate before hitting the ground. The precipitation can take the form of rain or ice particles but because it falls into dry air below the cloud it evaporates. Occasionally, because of their funnel-like shape, virga can be mistaken for tornadoes.

Monday

Date and Time .... Location ....

Relative Humidity ....   Wind Speed and Direction ....

Atmospheric Pressure ....   Cloud Cover..... oktas/overcast   Temperature ....

Cloud Types   ....

General Observations   ....

Predictions for Tomorrow   ....

Tuesday

Date and Time .... Location ....

Relative Humidity ....   Wind Speed and Direction ....

Atmospheric Pressure ....   Cloud Cover..... oktas/overcast   Temperature ....

Cloud Types   ....

General Observations   ....

Predictions for Tomorrow   ....

Wednesday

Date and Time .... Location ....

Relative Humidity ....   Wind Speed and Direction ....

Atmospheric Pressure ....   Cloud Cover..... oktas/overcast   Temperature ....

Cloud Types   ....

General Observations   ....

Predictions for Tomorrow   ....

Thursday

Date and Time .... Location ....

Relative Humidity ....   Wind Speed and Direction ....

Atmospheric Pressure ....   Cloud Cover..... oktas/overcast   Temperature ....

Cloud Types   ....

General Observations   ....

Predictions for Tomorrow   ....

Friday

Date and Time .... Location ....

Relative Humidity .... Wind Speed and Direction ....
Atmospheric Pressure .... Cloud Cover..... oktas/overcast Temperature ....

Cloud Types ....

General Observations ....

Predictions for Tomorrow ....

Saturday

Date and Time .... Location ....

Relative Humidity .... Wind Speed and Direction ....
Atmospheric Pressure .... Cloud Cover..... oktas/overcast Temperature ....

Cloud Types ....

General Observations ....

Predictions for Tomorrow ....

Sunday

Date and Time .... Location ....

Relative Humidity .... Wind Speed and Direction ....
Atmospheric Pressure .... Cloud Cover..... oktas/overcast Temperature ....

Cloud Types ....

General Observations ....

Predictions for Tomorrow ....

Weather Predictor's Tip

The shape of a raindrop is created by surface tension and air pressure. Raindrops tend to start out as round due to the attraction of water molecules to one another, but as they meet air resistance in their fall they undergo a change of shape. Since air pressure is greater on the bottom of the falling drop than on the sides, the undersides tend to flatten out.

Monday

Date and Time .... Location ....

Relative Humidity ....   Wind Speed and Direction ....

Atmospheric Pressure ....   Cloud Cover..... oktas/overcast   Temperature ....

Cloud Types ....

General Observations ....

Predictions for Tomorrow ....

Tuesday

Date and Time .... Location ....

Relative Humidity ....   Wind Speed and Direction ....

Atmospheric Pressure ....   Cloud Cover..... oktas/overcast   Temperature ....

Cloud Types ....

General Observations ....

Predictions for Tomorrow ....

Wednesday

Date and Time .... Location ....

Relative Humidity ....   Wind Speed and Direction ....

Atmospheric Pressure ....   Cloud Cover..... oktas/overcast   Temperature ....

Cloud Types ....

General Observations ....

Predictions for Tomorrow ....

Thursday

Date and Time .... Location ....

Relative Humidity ....   Wind Speed and Direction ....

Atmospheric Pressure ....   Cloud Cover..... oktas/overcast   Temperature ....

Cloud Types ....

General Observations ....

Predictions for Tomorrow ....

1029   1027

880

89

1029

Friday

Date
and Time .... Location ....

Relative Humidity .... Wind Speed and Direction ....

Atmospheric Pressure .... Cloud Cover..... oktas/overcast Temperature ....

Cloud Types ....

General Observations ....

Predictions for Tomorrow ....

Saturday

Date
and Time .... Location ....

Relative Humidity .... Wind Speed and Direction ....

Atmospheric Pressure .... Cloud Cover..... oktas/overcast Temperature ....

Cloud Types ....

General Observations ....

Predictions for Tomorrow ....

Sunday

Date
and Time .... Location ....

Relative Humidity .... Wind Speed and Direction ....

Atmospheric Pressure .... Cloud Cover..... oktas/overcast Temperature ....

Cloud Types ....

General Observations ....

Predictions for Tomorrow ....

Weather Predictor's Tip

When forecasters mention a probability of precipitation expressed as a percentage, they are referring to a combination of two figures: the likelihood of precipitation in a particular area and the percentage of the area that is likely to get it. So a 70 percent chance of rain means that there is a 70 percent chance of measurable precipitation in a specific area in a 12-hour period.

Monday

Date and Time .... Location ....

Relative Humidity ....   Wind Speed and Direction ....

Atmospheric Pressure ....   Cloud Cover..... oktas/overcast   Temperature ....

Cloud Types ....

General Observations ....

Predictions for Tomorrow ....

Tuesday

Date and Time .... Location ....

Relative Humidity ....   Wind Speed and Direction ....

Atmospheric Pressure ....   Cloud Cover..... oktas/overcast   Temperature ....

Cloud Types ....

General Observations ....

Predictions for Tomorrow ....

Wednesday

Date and Time .... Location ....

Relative Humidity ....   Wind Speed and Direction ....

Atmospheric Pressure ....   Cloud Cover..... oktas/overcast   Temperature ....

Cloud Types ....

General Observations ....

Predictions for Tomorrow ....

Thursday

Date and Time .... Location ....

Relative Humidity ....   Wind Speed and Direction ....

Atmospheric Pressure ....   Cloud Cover..... oktas/overcast   Temperature ....

Cloud Types ....

General Observations ....

Predictions for Tomorrow ....

1029   1027   880   89   1029

Friday

Date and Time .... Location ....

Relative Humidity .... Wind Speed and Direction ....

Atmospheric Pressure .... Cloud Cover..... oktas/overcast Temperature ....

Cloud Types ....

General Observations ....

Predictions for Tomorrow ....

Saturday

Date and Time .... Location ....

Relative Humidity .... Wind Speed and Direction ....

Atmospheric Pressure .... Cloud Cover..... oktas/overcast Temperature ....

Cloud Types ....

General Observations ....

Predictions for Tomorrow ....

Sunday

Date and Time .... Location ....

Relative Humidity .... Wind Speed and Direction ....

Atmospheric Pressure .... Cloud Cover..... oktas/overcast Temperature ....

Cloud Types ....

General Observations ....

Predictions for Tomorrow ....

Weather Predictor's Tip

Larger, severe thunderstorms tend to produce the most lightning bolts of the greatest intensity. Severe thunderstorms have powerful updrafts that can change the sound of thunder depending on the distance of the observer from the storm. At a distance of 6–12 miles (10–20 km) a severe storm will produce rolling thunder. Intense lightning bolts can create crackling sounds followed by a loud bang.

Monday

Date
and Time .... Location ....

Relative Humidity .... Wind Speed and Direction ....
Atmospheric Pressure .... Cloud Cover..... oktas/overcast Temperature ....

Cloud Types ....

General Observations ....

Predictions for Tomorrow ....

Tuesday

Date
and Time .... Location ....

Relative Humidity .... Wind Speed and Direction ....
Atmospheric Pressure .... Cloud Cover..... oktas/overcast Temperature ....

Cloud Types ....

General Observations ....

Predictions for Tomorrow ....

Wednesday

Date
and Time .... Location ....

Relative Humidity .... Wind Speed and Direction ....
Atmospheric Pressure .... Cloud Cover..... oktas/overcast Temperature ....

Cloud Types ....

General Observations ....

Predictions for Tomorrow ....

Thursday

Date
and Time .... Location ....

Relative Humidity .... Wind Speed and Direction ....
Atmospheric Pressure .... Cloud Cover..... oktas/overcast Temperature ....

Cloud Types ....

General Observations ....

Predictions for Tomorrow ....

Friday

Date
and Time .... Location ....

Relative Humidity ....   Wind Speed and Direction ....

Atmospheric Pressure ....   Cloud Cover..... oktas/overcast   Temperature ....

Cloud Types ....

General Observations ....

Predictions for Tomorrow ....

Saturday

Date
and Time .... Location ....

Relative Humidity ....   Wind Speed and Direction ....

Atmospheric Pressure ....   Cloud Cover..... oktas/overcast   Temperature ....

Cloud Types ....

General Observations ....

Predictions for Tomorrow ....

Sunday

Date
and Time .... Location ....

Relative Humidity ....   Wind Speed and Direction ....

Atmospheric Pressure ....   Cloud Cover..... oktas/overcast   Temperature ....

Cloud Types ....

General Observations ....

Predictions for Tomorrow ....

Weather Predictor's Tip

Air molecules will move faster and farther apart on a hot day than on a cool one, reducing air density. A decrease in air pressure will also decrease air density because the molecules will stay farther apart. The reduced air density is why a golf ball will travel farther on a course at higher elevations than at sea level.

Monday

Date
and Time .... Location ....

Relative Humidity ....   Wind Speed and Direction ....

Atmospheric Pressure ....   Cloud Cover..... oktas/overcast   Temperature ....

Cloud Types ....

General Observations ....

Predictions for Tomorrow ....

Tuesday

Date
and Time .... Location ....

Relative Humidity ....   Wind Speed and Direction ....

Atmospheric Pressure ....   Cloud Cover..... oktas/overcast   Temperature ....

Cloud Types ....

General Observations ....

Predictions for Tomorrow ....

Wednesday

Date
and Time .... Location ....

Relative Humidity ....   Wind Speed and Direction ....

Atmospheric Pressure ....   Cloud Cover..... oktas/overcast   Temperature ....

Cloud Types ....

General Observations ....

Predictions for Tomorrow ....

Thursday

Date
and Time .... Location ....

Relative Humidity ....   Wind Speed and Direction ....

Atmospheric Pressure ....   Cloud Cover..... oktas/overcast   Temperature ....

Cloud Types ....

General Observations ....

Predictions for Tomorrow ....

Friday

Date and Time .... Location ....

Relative Humidity .... Wind Speed and Direction ....
Atmospheric Pressure .... Cloud Cover..... oktas/overcast   Temperature ....

Cloud Types ....

General Observations ....

Predictions for Tomorrow ....

Saturday

Date and Time .... Location ....

Relative Humidity .... Wind Speed and Direction ....
Atmospheric Pressure .... Cloud Cover..... oktas/overcast   Temperature ....

Cloud Types ....

General Observations ....

Predictions for Tomorrow ....

Sunday

Date and Time .... Location ....

Relative Humidity .... Wind Speed and Direction ....
Atmospheric Pressure .... Cloud Cover..... oktas/overcast   Temperature ....

Cloud Types ....

General Observations ....

Predictions for Tomorrow ....

Weather Predictor's Tip

There is some truth to the old saying about the calm before the storm. Air inside the intense low-pressure center of a major storm will often rise and then descend outside the storm. This parcel of sinking air—high pressure—will bring clear skies and relatively light winds on the periphery of the storm, sometimes creating a "calm."

Monday

Date
and Time .... Location ....

Relative Humidity ....   Wind Speed and Direction ....

Atmospheric Pressure ....   Cloud Cover..... oktas/overcast   Temperature ....

Cloud Types ....

General Observations ....

Predictions for Tomorrow ....

Tuesday

Date
and Time .... Location ....

Relative Humidity ....   Wind Speed and Direction ....

Atmospheric Pressure ....   Cloud Cover..... oktas/overcast   Temperature ....

Cloud Types ....

General Observations ....

Predictions for Tomorrow ....

Wednesday

Date
and Time .... Location ....

Relative Humidity ....   Wind Speed and Direction ....

Atmospheric Pressure ....   Cloud Cover..... oktas/overcast   Temperature ....

Cloud Types ....

General Observations ....

Predictions for Tomorrow ....

Thursday

Date
and Time .... Location ....

Relative Humidity ....   Wind Speed and Direction ....

Atmospheric Pressure ....   Cloud Cover..... oktas/overcast   Temperature ....

Cloud Types ....

General Observations ....

Predictions for Tomorrow ....

Friday

Date
and Time .... Location ....

Relative Humidity ....   Wind Speed and Direction ....

Atmospheric Pressure ....   Cloud Cover..... oktas/overcast   Temperature ....

Cloud Types ....

General Observations ....

Predictions for Tomorrow ....

Saturday

Date
and Time .... Location ....

Relative Humidity ....   Wind Speed and Direction ....

Atmospheric Pressure ....   Cloud Cover..... oktas/overcast   Temperature ....

Cloud Types ....

General Observations ....

Predictions for Tomorrow ....

Sunday

Date
and Time .... Location ....

Relative Humidity ....   Wind Speed and Direction ....

Atmospheric Pressure ....   Cloud Cover..... oktas/overcast   Temperature ....

Cloud Types ....

General Observations ....

Predictions for Tomorrow ....

Weather Predictor's Tip

Generally, a car is a safe place to be during a thunderstorm because the metal body of the car acts as a lightning rod, gathering the electric field and protecting the interior. Talking on a telephone with a cord attached or being in contact with appliances, however, can be dangerous because the current may come through the wiring.

Monday

Date
and Time .... Location ....

Relative Humidity ....   Wind Speed and Direction ....

Atmospheric Pressure ....   Cloud Cover..... oktas/overcast   Temperature ....

Cloud Types ....

General Observations ....

Predictions for Tomorrow ....

Tuesday

Date
and Time .... Location ....

Relative Humidity ....   Wind Speed and Direction ....

Atmospheric Pressure ....   Cloud Cover..... oktas/overcast   Temperature ....

Cloud Types ....

General Observations ....

Predictions for Tomorrow ....

Wednesday

Date
and Time .... Location ....

Relative Humidity ....   Wind Speed and Direction ....

Atmospheric Pressure ....   Cloud Cover..... oktas/overcast   Temperature ....

Cloud Types ....

General Observations ....

Predictions for Tomorrow ....

Thursday

Date
and Time .... Location ....

Relative Humidity ....   Wind Speed and Direction ....

Atmospheric Pressure ....   Cloud Cover..... oktas/overcast   Temperature ....

Cloud Types ....

General Observations ....

Predictions for Tomorrow ....

Friday

Date
and Time .... Location ....

Relative Humidity ....   Wind Speed and Direction ....

Atmospheric Pressure ....   Cloud Cover..... oktas/overcast   Temperature ....

Cloud Types   ....

General Observations   ....

Predictions for Tomorrow   ....

Saturday

Date
and Time .... Location ....

Relative Humidity ....   Wind Speed and Direction ....

Atmospheric Pressure ....   Cloud Cover..... oktas/overcast   Temperature ....

Cloud Types   ....

General Observations   ....

Predictions for Tomorrow   ....

Sunday

Date
and Time .... Location ....

Relative Humidity ....   Wind Speed and Direction ....

Atmospheric Pressure ....   Cloud Cover..... oktas/overcast   Temperature ....

Cloud Types   ....

General Observations   ....

Predictions for Tomorrow   ....

Weather Predictor's Tip

A heat burst is a rare phenomenon that can occur if the air below a weakening thunderstorm is dry enough. As the precipitation evaporates, the air becomes cooler and denser than the air around it, which causes it to accelerate downward. If the air keeps sinking, it compresses, heats up, and may reach the ground as a "burst" of heat.

Monday

Date and Time .... Location ....

Relative Humidity ....   Wind Speed and Direction ....

Atmospheric Pressure ....   Cloud Cover..... oktas/overcast   Temperature ....

Cloud Types ....

General Observations ....

Predictions for Tomorrow ....

Tuesday

Date and Time .... Location ....

Relative Humidity ....   Wind Speed and Direction ....

Atmospheric Pressure ....   Cloud Cover..... oktas/overcast   Temperature ....

Cloud Types ....

General Observations ....

Predictions for Tomorrow ....

Wednesday

Date and Time .... Location ....

Relative Humidity ....   Wind Speed and Direction ....

Atmospheric Pressure ....   Cloud Cover..... oktas/overcast   Temperature ....

Cloud Types ....

General Observations ....

Predictions for Tomorrow ....

Thursday

Date and Time .... Location ....

Relative Humidity ....   Wind Speed and Direction ....

Atmospheric Pressure ....   Cloud Cover..... oktas/overcast   Temperature ....

Cloud Types ....

General Observations ....

Predictions for Tomorrow ....

Friday

Date and Time .... Location ....

Relative Humidity ....   Wind Speed and Direction ....

Atmospheric Pressure ....   Cloud Cover..... oktas/overcast   Temperature ....

Cloud Types ....

General Observations ....

Predictions for Tomorrow ....

Saturday

Date and Time .... Location ....

Relative Humidity ....   Wind Speed and Direction ....

Atmospheric Pressure ....   Cloud Cover..... oktas/overcast   Temperature ....

Cloud Types ....

General Observations ....

Predictions for Tomorrow ....

Sunday

Date and Time .... Location ....

Relative Humidity ....   Wind Speed and Direction ....

Atmospheric Pressure ....   Cloud Cover..... oktas/overcast   Temperature ....

Cloud Types ....

General Observations ....

Predictions for Tomorrow ....

Weather Predictor's Tip

So-called heat lightning is the same as regular lightning. People call it heat lightning because some bolts seem to be coming out of the blue on a hot day. What people are probably seeing is lightning from the top of the cumulonimbus cloud of a distant thunderstorm that cannot be discerned. Heat cannot produce lightning.

880

89

Monday

Date and Time .... Location ....

Relative Humidity .... Wind Speed and Direction ....
Atmospheric Pressure .... Cloud Cover..... oktas/overcast   Temperature ....

Cloud Types ....

General Observations ....

Predictions for Tomorrow ....

Tuesday

Date and Time .... Location ....

Relative Humidity .... Wind Speed and Direction ....
Atmospheric Pressure .... Cloud Cover..... oktas/overcast   Temperature ....

Cloud Types ....

General Observations ....

Predictions for Tomorrow ....

Wednesday

Date and Time .... Location ....

Relative Humidity .... Wind Speed and Direction ....
Atmospheric Pressure .... Cloud Cover..... oktas/overcast   Temperature ....

Cloud Types ....

General Observations ....

Predictions for Tomorrow ....

Thursday

Date and Time .... Location ....

Relative Humidity .... Wind Speed and Direction ....
Atmospheric Pressure .... Cloud Cover..... oktas/overcast   Temperature ....

Cloud Types ....

General Observations ....

Predictions for Tomorrow ....

Friday

Date and Time .... Location ....

Relative Humidity .... Wind Speed and Direction ....
Atmospheric Pressure .... Cloud Cover..... oktas/overcast Temperature ....

Cloud Types ....

General Observations ....

Predictions for Tomorrow ....

Saturday

Date and Time .... Location ....

Relative Humidity .... Wind Speed and Direction ....
Atmospheric Pressure .... Cloud Cover..... oktas/overcast Temperature ....

Cloud Types ....

General Observations ....

Predictions for Tomorrow ....

Sunday

Date and Time .... Location ....

Relative Humidity .... Wind Speed and Direction ....
Atmospheric Pressure .... Cloud Cover..... oktas/overcast Temperature ....

Cloud Types ....

General Observations ....

Predictions for Tomorrow ....

Weather Predictor's Tip

Precipitation falls mostly from the base of clouds and reaches the ground in a continuous vertical curtain. Typically, the rain cascade has a uniform, grayish shade or tinge. The motion of the rain cascade follows the motion of the cloud producing it. Even when several clouds are present, only one cloud type is usually responsible for the rain.

Monday

Date
and Time .... Location ....

Relative Humidity ....   Wind Speed and Direction ....

Atmospheric Pressure ....   Cloud Cover..... oktas/overcast   Temperature ....

Cloud Types ....

General Observations ....

Predictions for Tomorrow ....

Tuesday

Date
and Time .... Location ....

Relative Humidity ....   Wind Speed and Direction ....

Atmospheric Pressure ....   Cloud Cover..... oktas/overcast   Temperature ....

Cloud Types ....

General Observations ....

Predictions for Tomorrow ....

Wednesday

Date
and Time .... Location ....

Relative Humidity ....   Wind Speed and Direction ....

Atmospheric Pressure ....   Cloud Cover..... oktas/overcast   Temperature ....

Cloud Types ....

General Observations ....

Predictions for Tomorrow ....

Thursday

Date
and Time .... Location ....

Relative Humidity ....   Wind Speed and Direction ....

Atmospheric Pressure ....   Cloud Cover..... oktas/overcast   Temperature ....

Cloud Types ....

General Observations ....

Predictions for Tomorrow ....

Friday

Date and Time .... Location ....

Relative Humidity ....   Wind Speed and Direction ....

Atmospheric Pressure ....   Cloud Cover..... oktas/overcast   Temperature ....

Cloud Types ....

General Observations ....

Predictions for Tomorrow ....

Saturday

Date and Time .... Location ....

Relative Humidity ....   Wind Speed and Direction ....

Atmospheric Pressure ....   Cloud Cover..... oktas/overcast   Temperature ....

Cloud Types ....

General Observations ....

Predictions for Tomorrow ....

Sunday

Date and Time .... Location ....

Relative Humidity ....   Wind Speed and Direction ....

Atmospheric Pressure ....   Cloud Cover..... oktas/overcast   Temperature ....

Cloud Types ....

General Observations ....

Predictions for Tomorrow ....

Weather Predictor's Tip

Interpreting the meteorological significance of wind information is difficult and should be done in conjunction with other data, such as temperature, humidity, and barometric pressure. Strong winds indicate unstable conditions, while light to calm winds typically indicate the atmosphere is stable. Wind information, combined with other data regularly monitored, indicates how the atmosphere's stability is changing over time.

Monday

Date
and Time .... Location ....

Relative Humidity .... Wind Speed and Direction ....
Atmospheric Pressure .... Cloud Cover..... oktas/overcast Temperature ....

Cloud Types ....

General Observations ....

Predictions for Tomorrow ....

Tuesday

Date
and Time .... Location ....

Relative Humidity .... Wind Speed and Direction ....
Atmospheric Pressure .... Cloud Cover..... oktas/overcast Temperature ....

Cloud Types ....

General Observations ....

Predictions for Tomorrow ....

Wednesday

Date
and Time .... Location ....

Relative Humidity .... Wind Speed and Direction ....
Atmospheric Pressure .... Cloud Cover..... oktas/overcast Temperature ....

Cloud Types ....

General Observations ....

Predictions for Tomorrow ....

Thursday

Date
and Time .... Location ....

Relative Humidity .... Wind Speed and Direction ....
Atmospheric Pressure .... Cloud Cover..... oktas/overcast Temperature ....

Cloud Types ....

General Observations ....

Predictions for Tomorrow ....

Friday

Date
and Time .... Location ....

Relative Humidity .... Wind Speed and Direction ....
Atmospheric Pressure .... Cloud Cover..... oktas/overcast Temperature ....

Cloud Types ....

General Observations ....

Predictions for Tomorrow ....

Saturday

Date
and Time .... Location ....

Relative Humidity .... Wind Speed and Direction ....
Atmospheric Pressure .... Cloud Cover..... oktas/overcast Temperature ....

Cloud Types ....

General Observations ....

Predictions for Tomorrow ....

Sunday

Date
and Time .... Location ....

Relative Humidity .... Wind Speed and Direction ....
Atmospheric Pressure .... Cloud Cover..... oktas/overcast Temperature ....

Cloud Types ....

General Observations ....

Predictions for Tomorrow ....

Weather Predictor's Tip

Weather instruments should be protected from direct sunlight and precipitation.
Instrument shelters can be crude and made from a white-painted box, or even
a short section of PVC drainpipe. Or you may prefer to purchase a Stevenson
screen. Weather stations should be placed in an open space, clear of trees
and buildings.

Monday

Date
and Time .... Location ....

Relative Humidity .... Wind Speed and Direction ....
Atmospheric Pressure .... Cloud Cover..... oktas/overcast Temperature ....

Cloud Types ....

General Observations ....

Predictions for Tomorrow ....

Tuesday

Date
and Time .... Location ....

Relative Humidity .... Wind Speed and Direction ....
Atmospheric Pressure .... Cloud Cover..... oktas/overcast Temperature ....

Cloud Types ....

General Observations ....

Predictions for Tomorrow ....

Wednesday

Date
and Time .... Location ....

Relative Humidity .... Wind Speed and Direction ....
Atmospheric Pressure .... Cloud Cover..... oktas/overcast Temperature ....

Cloud Types ....

General Observations ....

Predictions for Tomorrow ....

Thursday

Date
and Time .... Location ....

Relative Humidity .... Wind Speed and Direction ....
Atmospheric Pressure .... Cloud Cover..... oktas/overcast Temperature ....

Cloud Types ....

General Observations ....

Predictions for Tomorrow ....

1029 1027

880

89

1029

Friday

Date and Time .... Location ....

Relative Humidity ....   Wind Speed and Direction ....

Atmospheric Pressure ....   Cloud Cover..... oktas/overcast   Temperature ....

Cloud Types ....

General Observations ....

Predictions for Tomorrow ....

Saturday

Date and Time .... Location ....

Relative Humidity ....   Wind Speed and Direction ....

Atmospheric Pressure ....   Cloud Cover..... oktas/overcast   Temperature ....

Cloud Types ....

General Observations ....

Predictions for Tomorrow ....

Sunday

Date and Time .... Location ....

Relative Humidity ....   Wind Speed and Direction ....

Atmospheric Pressure ....   Cloud Cover..... oktas/overcast   Temperature ....

Cloud Types ....

General Observations ....

Predictions for Tomorrow ....

Weather Predictor's Tip

The greater the depth of a rain band, the less background light it allows to penetrate, so the darker it should look to the observer. Rain bands do not necessarily approach by moving perpendicular to your line of sight but tend to arrive at an angle. To determine when rain will reach your area, you should observe the movement of the parent cloud responsible for the precipitation.

Monday

Date and Time .... Location ....

Relative Humidity ....   Wind Speed and Direction ....

Atmospheric Pressure ....   Cloud Cover..... oktas/overcast   Temperature ....

Cloud Types ....

General Observations ....

Predictions for Tomorrow ....

Tuesday

Date and Time .... Location ....

Relative Humidity ....   Wind Speed and Direction ....

Atmospheric Pressure ....   Cloud Cover..... oktas/overcast   Temperature ....

Cloud Types ....

General Observations ....

Predictions for Tomorrow ....

Wednesday

Date and Time .... Location ....

Relative Humidity ....   Wind Speed and Direction ....

Atmospheric Pressure ....   Cloud Cover..... oktas/overcast   Temperature ....

Cloud Types ....

General Observations ....

Predictions for Tomorrow ....

Thursday

Date and Time .... Location ....

Relative Humidity ....   Wind Speed and Direction ....

Atmospheric Pressure ....   Cloud Cover..... oktas/overcast   Temperature ....

Cloud Types ....

General Observations ....

Predictions for Tomorrow ....

Friday

Date and Time .... Location ....

Relative Humidity ....   Wind Speed and Direction ....

Atmospheric Pressure ....   Cloud Cover..... oktas/overcast   Temperature ....

Cloud Types   ....

General Observations   ....

Predictions for Tomorrow   ....

Saturday

Date and Time .... Location ....

Relative Humidity ....   Wind Speed and Direction ....

Atmospheric Pressure ....   Cloud Cover..... oktas/overcast   Temperature ....

Cloud Types   ....

General Observations   ....

Predictions for Tomorrow   ....

Sunday

Date and Time .... Location ....

Relative Humidity ....   Wind Speed and Direction ....

Atmospheric Pressure ....   Cloud Cover..... oktas/overcast   Temperature ....

Cloud Types   ....

General Observations   ....

Predictions for Tomorrow   ....

Weather Predictor's Tip

The size of a tornado is not necessarily an indication of how intense it is. A large tornado can be weak, as seven out of ten tornados are, while a small tornado may be much more violent. The intensity of a tornado is measured by the Fujita scale, based on the damage caused after it has passed over a manmade structure.

Monday

Date
and Time .... Location ....

Relative Humidity .... Wind Speed and Direction ....
Atmospheric Pressure .... Cloud Cover..... oktas/overcast Temperature ....

Cloud Types ....

General Observations ....

Predictions for Tomorrow ....

Tuesday

Date
and Time .... Location ....

Relative Humidity .... Wind Speed and Direction ....
Atmospheric Pressure .... Cloud Cover..... oktas/overcast Temperature ....

Cloud Types ....

General Observations ....

Predictions for Tomorrow ....

Wednesday

Date
and Time .... Location ....

Relative Humidity .... Wind Speed and Direction ....
Atmospheric Pressure .... Cloud Cover..... oktas/overcast Temperature ....

Cloud Types ....

General Observations ....

Predictions for Tomorrow ....

Thursday

Date
and Time .... Location ....

Relative Humidity .... Wind Speed and Direction ....
Atmospheric Pressure .... Cloud Cover..... oktas/overcast Temperature ....

Cloud Types ....

General Observations ....

Predictions for Tomorrow ....

Friday

Date and Time .... Location ....

Relative Humidity ....   Wind Speed and Direction ....

Atmospheric Pressure ....   Cloud Cover..... oktas/overcast   Temperature ....

Cloud Types   ....

General Observations   ....

Predictions for Tomorrow   ....

Saturday

Date and Time .... Location ....

Relative Humidity ....   Wind Speed and Direction ....

Atmospheric Pressure ....   Cloud Cover..... oktas/overcast   Temperature ....

Cloud Types   ....

General Observations   ....

Predictions for Tomorrow   ....

Sunday

Date and Time .... Location ....

Relative Humidity ....   Wind Speed and Direction ....

Atmospheric Pressure ....   Cloud Cover..... oktas/overcast   Temperature ....

Cloud Types   ....

General Observations   ....

Predictions for Tomorrow   ....

Weather Predictor's Tip

Cloud motions vary according to height and air flows in the upper atmosphere. To observe a cloud's motion you should concentrate on one cloud, remain as still as possible, and have a compass to determine its direction, using a landmark to locate your position on a map. In cases of slow-moving clouds associated with thunderstorm development, you can measure their motion at five-minute intervals.

Monday

Date and Time .... Location ....

Relative Humidity .... Wind Speed and Direction ....

Atmospheric Pressure .... Cloud Cover..... oktas/overcast   Temperature ....

Cloud Types ....

General Observations ....

Predictions for Tomorrow ....

Tuesday

Date and Time .... Location ....

Relative Humidity .... Wind Speed and Direction ....

Atmospheric Pressure .... Cloud Cover..... oktas/overcast   Temperature ....

Cloud Types ....

General Observations ....

Predictions for Tomorrow ....

Wednesday

Date and Time .... Location ....

Relative Humidity .... Wind Speed and Direction ....

Atmospheric Pressure .... Cloud Cover..... oktas/overcast   Temperature ....

Cloud Types ....

General Observations ....

Predictions for Tomorrow ....

Thursday

Date and Time .... Location ....

Relative Humidity .... Wind Speed and Direction ....

Atmospheric Pressure .... Cloud Cover..... oktas/overcast   Temperature ....

Cloud Types ....

General Observations ....

Predictions for Tomorrow ....

Friday

Date and Time .... Location ....

Relative Humidity ....   Wind Speed and Direction ....

Atmospheric Pressure ....   Cloud Cover..... oktas/overcast   Temperature ....

Cloud Types ....

General Observations ....

Predictions for Tomorrow ....

Saturday

Date and Time .... Location ....

Relative Humidity ....   Wind Speed and Direction ....

Atmospheric Pressure ....   Cloud Cover..... oktas/overcast   Temperature ....

Cloud Types ....

General Observations ....

Predictions for Tomorrow ....

Sunday

Date and Time .... Location ....

Relative Humidity ....   Wind Speed and Direction ....

Atmospheric Pressure ....   Cloud Cover..... oktas/overcast   Temperature ....

Cloud Types ....

General Observations ....

Predictions for Tomorrow ....

Weather Predictor's Tip

Cape Verde-type hurricanes are storms that have developed in the Atlantic Ocean close to the Cape Verde Islands and become hurricanes before reaching the Caribbean. They typically occur in August and September (hurricane season begins in June and lasts until November) and average about two every hurricane season.

Monday

Date and Time .... Location ....

Relative Humidity .... Wind Speed and Direction ....
Atmospheric Pressure .... Cloud Cover..... oktas/overcast   Temperature ....

Cloud Types ....

General Observations ....

Predictions for Tomorrow ....

Tuesday

Date and Time .... Location ....

Relative Humidity .... Wind Speed and Direction ....
Atmospheric Pressure .... Cloud Cover..... oktas/overcast   Temperature ....

Cloud Types ....

General Observations ....

Predictions for Tomorrow ....

Wednesday

Date and Time .... Location ....

Relative Humidity .... Wind Speed and Direction ....
Atmospheric Pressure .... Cloud Cover..... oktas/overcast   Temperature ....

Cloud Types ....

General Observations ....

Predictions for Tomorrow ....

Thursday

Date and Time .... Location ....

Relative Humidity .... Wind Speed and Direction ....
Atmospheric Pressure .... Cloud Cover..... oktas/overcast   Temperature ....

Cloud Types ....

General Observations ....

Predictions for Tomorrow ....

Friday

Date and Time .... Location ....

Relative Humidity ....   Wind Speed and Direction ....
Atmospheric Pressure ....   Cloud Cover..... oktas/overcast   Temperature ....

Cloud Types ....

General Observations ....

Predictions for Tomorrow ....

Saturday

Date and Time .... Location ....

Relative Humidity ....   Wind Speed and Direction ....
Atmospheric Pressure ....   Cloud Cover..... oktas/overcast   Temperature ....

Cloud Types ....

General Observations ....

Predictions for Tomorrow ....

Sunday

Date and Time .... Location ....

Relative Humidity ....   Wind Speed and Direction ....
Atmospheric Pressure ....   Cloud Cover..... oktas/overcast   Temperature ....

Cloud Types ....

General Observations ....

Predictions for Tomorrow ....

Weather Predictor's Tip

When a weather service issues a "watch" it means that conditions favor the development of a serious weather event such as a storm or high winds. It warrants attention but it does not invariably mean that the event will occur. By contrast, a "warning" means that the event is impending or is already in progress.

Monday

Date and Time .... Location ....

Relative Humidity .... Wind Speed and Direction ....
Atmospheric Pressure .... Cloud Cover..... oktas/overcast   Temperature ....

Cloud Types ....

General Observations ....

Predictions for Tomorrow ....

Tuesday

Date and Time .... Location ....

Relative Humidity .... Wind Speed and Direction ....
Atmospheric Pressure .... Cloud Cover..... oktas/overcast   Temperature ....

Cloud Types ....

General Observations ....

Predictions for Tomorrow ....

Wednesday

Date and Time .... Location ....

Relative Humidity .... Wind Speed and Direction ....
Atmospheric Pressure .... Cloud Cover..... oktas/overcast   Temperature ....

Cloud Types ....

General Observations ....

Predictions for Tomorrow ....

Thursday

Date and Time .... Location ....

Relative Humidity .... Wind Speed and Direction ....
Atmospheric Pressure .... Cloud Cover..... oktas/overcast   Temperature ....

Cloud Types ....

General Observations ....

Predictions for Tomorrow ....

Friday

Date
and Time .... Location ....

Relative Humidity ....   Wind Speed and Direction ....

Atmospheric Pressure ....   Cloud Cover..... oktas/overcast   Temperature ....

Cloud Types   ....

General Observations   ....

Predictions for Tomorrow   ....

Saturday

Date
and Time .... Location ....

Relative Humidity ....   Wind Speed and Direction ....

Atmospheric Pressure ....   Cloud Cover..... oktas/overcast   Temperature ....

Cloud Types   ....

General Observations   ....

Predictions for Tomorrow   ....

Sunday

Date
and Time .... Location ....

Relative Humidity ....   Wind Speed and Direction ....

Atmospheric Pressure ....   Cloud Cover..... oktas/overcast   Temperature ....

Cloud Types   ....

General Observations   ....

Predictions for Tomorrow   ....

Weather Predictor's Tip

A "major hurricane" is a hurricane that can reach maximum sustained
surface winds for one minute of at least 111 mph (177 km/h). This is the
equivalent of a category 3, 4, or 5 hurricane on the Saffir-Simpson scale.
Major hurricanes—the official term—are also known, unofficially, as
intense hurricanes.

Monday

Date and Time .... Location ....

Relative Humidity ....   Wind Speed and Direction ....

Atmospheric Pressure ....   Cloud Cover..... oktas/overcast   Temperature ....

Cloud Types ....

General Observations ....

Predictions for Tomorrow ....

Tuesday

Date and Time .... Location ....

Relative Humidity ....   Wind Speed and Direction ....

Atmospheric Pressure ....   Cloud Cover..... oktas/overcast   Temperature ....

Cloud Types ....

General Observations ....

Predictions for Tomorrow ....

Wednesday

Date and Time .... Location ....

Relative Humidity ....   Wind Speed and Direction ....

Atmospheric Pressure ....   Cloud Cover..... oktas/overcast   Temperature ....

Cloud Types ....

General Observations ....

Predictions for Tomorrow ....

Thursday

Date and Time .... Location ....

Relative Humidity ....   Wind Speed and Direction ....

Atmospheric Pressure ....   Cloud Cover..... oktas/overcast   Temperature ....

Cloud Types ....

General Observations ....

Predictions for Tomorrow ....

Friday

Date and Time .... Location ....

Relative Humidity ....   Wind Speed and Direction ....

Atmospheric Pressure ....   Cloud Cover..... oktas/overcast   Temperature ....

Cloud Types ....

General Observations ....

Predictions for Tomorrow ....

Saturday

Date and Time .... Location ....

Relative Humidity ....   Wind Speed and Direction ....

Atmospheric Pressure ....   Cloud Cover..... oktas/overcast   Temperature ....

Cloud Types ....

General Observations ....

Predictions for Tomorrow ....

Sunday

Date and Time .... Location ....

Relative Humidity ....   Wind Speed and Direction ....

Atmospheric Pressure ....   Cloud Cover..... oktas/overcast   Temperature ....

Cloud Types ....

General Observations ....

Predictions for Tomorrow ....

Weather Predictor's Tip

For the best results, you should set your barometer on a day with anticyclonic conditions—little wind and clear and sunny skies—because the pressure will not change too much. To confirm that the reading is correct, you should check with the barometric reading for your location given by your national weather service.

Monday

Date
and Time .... Location ....

880

Relative Humidity ....   Wind Speed and Direction ....

89

Atmospheric Pressure ....   Cloud Cover..... oktas/overcast   Temperature ....

Cloud Types ....

General Observations ....

Predictions for Tomorrow ....

Tuesday

Date
and Time .... Location ....

Relative Humidity ....   Wind Speed and Direction ....

Atmospheric Pressure ....   Cloud Cover..... oktas/overcast   Temperature ....

Cloud Types ....

General Observations ....

Predictions for Tomorrow ....

Wednesday

Date
and Time .... Location ....

Relative Humidity ....   Wind Speed and Direction ....

Atmospheric Pressure ....   Cloud Cover..... oktas/overcast   Temperature ....

Cloud Types ....

General Observations ....

Predictions for Tomorrow ....

Thursday

Date
and Time .... Location ....

Relative Humidity ....   Wind Speed and Direction ....

Atmospheric Pressure ....   Cloud Cover..... oktas/overcast   Temperature ....

Cloud Types ....

General Observations ....

Predictions for Tomorrow ....

Friday

Date
and Time .... Location ....

Relative Humidity ....   Wind Speed and Direction ....

Atmospheric Pressure ....   Cloud Cover ..... oktas/overcast   Temperature ....

Cloud Types ....

General Observations ....

Predictions for Tomorrow ....

Saturday

Date
and Time .... Location ....

Relative Humidity ....   Wind Speed and Direction ....

Atmospheric Pressure ....   Cloud Cover ..... oktas/overcast   Temperature ....

Cloud Types ....

General Observations ....

Predictions for Tomorrow ....

Sunday

Date
and Time .... Location ....

Relative Humidity ....   Wind Speed and Direction ....

Atmospheric Pressure ....   Cloud Cover ..... oktas/overcast   Temperature ....

Cloud Types ....

General Observations ....

Predictions for Tomorrow ....

Weather Predictor's Tip

A landspout is similar to a waterspout, but over land; both are actually small and weak tornadoes. Like a tornado, a landspout is a rotating column of air that extends from a cumulonimbus cloud to the ground. It is usually associated with the outward flow of air from a thunderstorm. Landspouts often appear first on the ground in the form of a debris cloud. Then as the tornado matures, the debris is pulled upward.

Monday

Date and Time .... Location ....

Relative Humidity ....   Wind Speed and Direction ....

Atmospheric Pressure ....   Cloud Cover..... oktas/overcast   Temperature ....

Cloud Types ....

General Observations ....

Predictions for Tomorrow ....

Tuesday

Date and Time .... Location ....

Relative Humidity ....   Wind Speed and Direction ....

Atmospheric Pressure ....   Cloud Cover..... oktas/overcast   Temperature ....

Cloud Types ....

General Observations ....

Predictions for Tomorrow ....

Wednesday

Date and Time .... Location ....

Relative Humidity ....   Wind Speed and Direction ....

Atmospheric Pressure ....   Cloud Cover..... oktas/overcast   Temperature ....

Cloud Types ....

General Observations ....

Predictions for Tomorrow ....

Thursday

Date and Time .... Location ....

Relative Humidity ....   Wind Speed and Direction ....

Atmospheric Pressure ....   Cloud Cover..... oktas/overcast   Temperature ....

Cloud Types ....

General Observations ....

Predictions for Tomorrow ....

Friday

Date and Time .... Location ....

Relative Humidity ....   Wind Speed and Direction ....

Atmospheric Pressure ....   Cloud Cover..... oktas/overcast   Temperature ....

Cloud Types ....

General Observations ....

Predictions for Tomorrow ....

Saturday

Date and Time .... Location ....

Relative Humidity ....   Wind Speed and Direction ....

Atmospheric Pressure ....   Cloud Cover..... oktas/overcast   Temperature ....

Cloud Types ....

General Observations ....

Predictions for Tomorrow ....

Sunday

Date and Time .... Location ....

Relative Humidity ....   Wind Speed and Direction ....

Atmospheric Pressure ....   Cloud Cover..... oktas/overcast   Temperature ....

Cloud Types ....

General Observations ....

Predictions for Tomorrow ....

Weather Predictor's Tip

There is a distinction between rain and showers. The term "rain" is
used to refer to widespread rain, or rain that occurs over a large area
for a sustained period of time. "Showers" are more likely to be brief and
sporadic in duration—from a minute to an hour—and concentrated in only
limited locations.

Monday

Date and Time .... Location ....

Relative Humidity .... Wind Speed and Direction ....

Atmospheric Pressure .... Cloud Cover..... oktas/overcast   Temperature ....

Cloud Types ....

General Observations ....

Predictions for Tomorrow ....

Tuesday

Date and Time .... Location ....

Relative Humidity .... Wind Speed and Direction ....

Atmospheric Pressure .... Cloud Cover..... oktas/overcast   Temperature ....

Cloud Types ....

General Observations ....

Predictions for Tomorrow ....

Wednesday

Date and Time .... Location ....

Relative Humidity .... Wind Speed and Direction ....

Atmospheric Pressure .... Cloud Cover..... oktas/overcast   Temperature ....

Cloud Types ....

General Observations ....

Predictions for Tomorrow ....

Thursday

Date and Time .... Location ....

Relative Humidity .... Wind Speed and Direction ....

Atmospheric Pressure .... Cloud Cover..... oktas/overcast   Temperature ....

Cloud Types ....

General Observations ....

Predictions for Tomorrow ....

Friday

Date
and Time .... Location ....

Relative Humidity .... Wind Speed and Direction ....
Atmospheric Pressure .... Cloud Cover..... oktas/overcast Temperature ....

Cloud Types ....

General Observations ....

Predictions for Tomorrow ....

Saturday

Date
and Time .... Location ....

Relative Humidity .... Wind Speed and Direction ....
Atmospheric Pressure .... Cloud Cover..... oktas/overcast Temperature ....

Cloud Types ....

General Observations ....

Predictions for Tomorrow ....

Sunday

Date
and Time .... Location ....

Relative Humidity .... Wind Speed and Direction ....
Atmospheric Pressure .... Cloud Cover..... oktas/overcast Temperature ....

Cloud Types ....

General Observations ....

Predictions for Tomorrow ....

Weather Predictor's Tip

Funnel clouds are generally considered to be tornadoes that do not reach the
ground, but in certain regions, such as the northwest United States, what appear to
be funnel clouds are actually a tightly wrapped rush of rising air that can assume a
funnel shape. They are known as cold-air funnels and do not bring the destruction
that a tornado does.

Monday

Date and Time .... Location ....

Relative Humidity ....   Wind Speed and Direction ....

Atmospheric Pressure ....   Cloud Cover..... oktas/overcast   Temperature ....

Cloud Types   ....

General Observations   ....

Predictions for Tomorrow   ....

Tuesday

Date and Time .... Location ....

Relative Humidity ....   Wind Speed and Direction ....

Atmospheric Pressure ....   Cloud Cover..... oktas/overcast   Temperature ....

Cloud Types   ....

General Observations   ....

Predictions for Tomorrow   ....

Wednesday

Date and Time .... Location ....

Relative Humidity ....   Wind Speed and Direction ....

Atmospheric Pressure ....   Cloud Cover..... oktas/overcast   Temperature ....

Cloud Types   ....

General Observations   ....

Predictions for Tomorrow   ....

Thursday

Date and Time .... Location ....

Relative Humidity ....   Wind Speed and Direction ....

Atmospheric Pressure ....   Cloud Cover..... oktas/overcast   Temperature ....

Cloud Types   ....

General Observations   ....

Predictions for Tomorrow   ....

Friday

Date and Time .... Location ....

Relative Humidity .... Wind Speed and Direction ....
Atmospheric Pressure .... Cloud Cover..... oktas/overcast   Temperature ....

Cloud Types ....

General Observations ....

Predictions for Tomorrow ....

Saturday

Date and Time .... Location ....

Relative Humidity .... Wind Speed and Direction ....
Atmospheric Pressure .... Cloud Cover..... oktas/overcast   Temperature ....

Cloud Types ....

General Observations ....

Predictions for Tomorrow ....

Sunday

Date and Time .... Location ....

Relative Humidity .... Wind Speed and Direction ....
Atmospheric Pressure .... Cloud Cover..... oktas/overcast   Temperature ....

Cloud Types ....

General Observations ....

Predictions for Tomorrow ....

Weather Predictor's Tip

High- and low-pressure readings do not necessarily forecast temperature. High pressure can bring very warm weather or very cold weather. It depends on where the air mass is coming from. Similarly, low pressure is often found near fronts, where air of warm and cold temperatures can exist in close proximity.

880

890

Monday

Date
and Time .... Location ....

Relative Humidity .... Wind Speed and Direction ....
Atmospheric Pressure .... Cloud Cover..... oktas/overcast Temperature ....

Cloud Types   ....

General Observations ....

Predictions for Tomorrow ....

Tuesday

Date
and Time .... Location ....

Relative Humidity .... Wind Speed and Direction ....
Atmospheric Pressure .... Cloud Cover..... oktas/overcast Temperature ....

Cloud Types   ....

General Observations ....

Predictions for Tomorrow ....

Wednesday

Date
and Time .... Location ....

Relative Humidity .... Wind Speed and Direction ....
Atmospheric Pressure .... Cloud Cover..... oktas/overcast Temperature ....

Cloud Types   ....

General Observations ....

Predictions for Tomorrow ....

Thursday

Date
and Time .... Location ....

Relative Humidity .... Wind Speed and Direction ....
Atmospheric Pressure .... Cloud Cover..... oktas/overcast Temperature ....

Cloud Types   ....

General Observations ....

Predictions for Tomorrow ....

Friday

Date
and Time .... Location ....

Relative Humidity ....   Wind Speed and Direction ....

Atmospheric Pressure ....   Cloud Cover..... oktas/overcast   Temperature ....

Cloud Types ....

General Observations ....

Predictions for Tomorrow ....

Saturday

Date
and Time .... Location ....

Relative Humidity ....   Wind Speed and Direction ....

Atmospheric Pressure ....   Cloud Cover..... oktas/overcast   Temperature ....

Cloud Types ....

General Observations ....

Predictions for Tomorrow ....

Sunday

Date
and Time .... Location ....

Relative Humidity ....   Wind Speed and Direction ....

Atmospheric Pressure ....   Cloud Cover..... oktas/overcast   Temperature ....

Cloud Types ....

General Observations ....

Predictions for Tomorrow ....

Weather Predictor's Tip

It is possible to estimate wind speed even without instruments: rustling leaves
indicate a 5–10 mph (8–16 km/h) breeze; branches begin to move around
15 mph (24 km/h); an umbrella will turn inside out when the wind reaches 30 mph
(48 km/h); branches snap at 40 mph (64 km/h); trees will be at risk of toppling when
wind reaches 60 mph (96 km/h).

Monday

Date and Time .... Location ....

Relative Humidity ....   Wind Speed and Direction ....

Atmospheric Pressure ....   Cloud Cover..... oktas/overcast   Temperature ....

Cloud Types ....

General Observations ....

Predictions for Tomorrow ....

Tuesday

Date and Time .... Location ....

Relative Humidity ....   Wind Speed and Direction ....

Atmospheric Pressure ....   Cloud Cover..... oktas/overcast   Temperature ....

Cloud Types ....

General Observations ....

Predictions for Tomorrow ....

Wednesday

Date and Time .... Location ....

Relative Humidity ....   Wind Speed and Direction ....

Atmospheric Pressure ....   Cloud Cover..... oktas/overcast   Temperature ....

Cloud Types ....

General Observations ....

Predictions for Tomorrow ....

Thursday

Date and Time .... Location ....

Relative Humidity ....   Wind Speed and Direction ....

Atmospheric Pressure ....   Cloud Cover..... oktas/overcast   Temperature ....

Cloud Types ....

General Observations ....

Predictions for Tomorrow ....

Friday

Date and Time .... Location ....

Relative Humidity .... Wind Speed and Direction ....

Atmospheric Pressure .... Cloud Cover..... oktas/overcast Temperature ....

Cloud Types ....

General Observations ....

Predictions for Tomorrow ....

Saturday

Date and Time .... Location ....

Relative Humidity .... Wind Speed and Direction ....

Atmospheric Pressure .... Cloud Cover..... oktas/overcast Temperature ....

Cloud Types ....

General Observations ....

Predictions for Tomorrow ....

Sunday

Date and Time .... Location ....

Relative Humidity .... Wind Speed and Direction ....

Atmospheric Pressure .... Cloud Cover..... oktas/overcast Temperature ....

Cloud Types ....

General Observations ....

Predictions for Tomorrow ....

Weather Predictor's Tip

The relative humidity refers to how much water vapor a volume of air contains relative to the maximum it could contain at that temperature. So air that is at 80 percent relative humidity contains 80 percent of the water vapor it could hold. At 100 percent the air is said to be saturated. For rain to start, the air only has to reach 100 percent relative humidity where it is raining.

Monday

Date and Time .... Location ....

Relative Humidity ....   Wind Speed and Direction ....

Atmospheric Pressure ....   Cloud Cover..... oktas/overcast   Temperature ....

Cloud Types ....

General Observations ....

Predictions for Tomorrow ....

Tuesday

Date and Time .... Location ....

Relative Humidity ....   Wind Speed and Direction ....

Atmospheric Pressure ....   Cloud Cover..... oktas/overcast   Temperature ....

Cloud Types ....

General Observations ....

Predictions for Tomorrow ....

Wednesday

Date and Time .... Location ....

Relative Humidity ....   Wind Speed and Direction ....

Atmospheric Pressure ....   Cloud Cover..... oktas/overcast   Temperature ....

Cloud Types ....

General Observations ....

Predictions for Tomorrow ....

Thursday

Date and Time .... Location ....

Relative Humidity ....   Wind Speed and Direction ....

Atmospheric Pressure ....   Cloud Cover..... oktas/overcast   Temperature ....

Cloud Types ....

General Observations ....

Predictions for Tomorrow ....

Friday

Date and Time .... Location ....

Relative Humidity .... Wind Speed and Direction ....

Atmospheric Pressure .... Cloud Cover..... oktas/overcast Temperature ....

Cloud Types ....

General Observations ....

Predictions for Tomorrow ....

Saturday

Date and Time .... Location ....

Relative Humidity .... Wind Speed and Direction ....

Atmospheric Pressure .... Cloud Cover..... oktas/overcast Temperature ....

Cloud Types ....

General Observations ....

Predictions for Tomorrow ....

Sunday

Date and Time .... Location ....

Relative Humidity .... Wind Speed and Direction ....

Atmospheric Pressure .... Cloud Cover..... oktas/overcast Temperature ....

Cloud Types ....

General Observations ....

Predictions for Tomorrow ....

Weather Predictor's Tip

"Tornado alley" is a region in the United States that experiences more tornadoes than anywhere else in the world. It includes portions of Texas, Oklahoma, Kansas, and Nebraska. This region gets warm, moist air from the Gulf of Mexico, cold air from Canada, and dry air from the west—all of which causes the air to become unstable and fertile for the formation of tornadoes.

Monday

Date
and Time .... Location ....

Relative Humidity ....   Wind Speed and Direction ....
Atmospheric Pressure ....   Cloud Cover..... oktas/overcast   Temperature ....

Cloud Types   ....

General Observations   ....

Predictions for Tomorrow   ....

Tuesday

Date
and Time .... Location ....

Relative Humidity ....   Wind Speed and Direction ....
Atmospheric Pressure ....   Cloud Cover..... oktas/overcast   Temperature ....

Cloud Types   ....

General Observations   ....

Predictions for Tomorrow   ....

Wednesday

Date
and Time .... Location ....

Relative Humidity ....   Wind Speed and Direction ....
Atmospheric Pressure ....   Cloud Cover..... oktas/overcast   Temperature ....

Cloud Types   ....

General Observations   ....

Predictions for Tomorrow   ....

Thursday

Date
and Time .... Location ....

Relative Humidity ....   Wind Speed and Direction ....
Atmospheric Pressure ....   Cloud Cover..... oktas/overcast   Temperature ....

Cloud Types   ....

General Observations   ....

Predictions for Tomorrow   ....

Friday

Date
and Time .... Location ....

Relative Humidity .... Wind Speed and Direction ....

Atmospheric Pressure .... Cloud Cover..... oktas/overcast Temperature ....

Cloud Types ....

General Observations ....

Predictions for Tomorrow ....

Saturday

Date
and Time .... Location ....

Relative Humidity .... Wind Speed and Direction ....

Atmospheric Pressure .... Cloud Cover..... oktas/overcast Temperature ....

Cloud Types ....

General Observations ....

Predictions for Tomorrow ....

Sunday

Date
and Time .... Location ....

Relative Humidity .... Wind Speed and Direction ....

Atmospheric Pressure .... Cloud Cover..... oktas/overcast Temperature ....

Cloud Types ....

General Observations ....

Predictions for Tomorrow ....

Weather Predictor's Tip

Temperatures will be lower when the ground is covered with snow because snow reflects much of the Sun's energy back into space without its being absorbed by the ground. Conversely, in winters with little snow cover, temperatures will be warmer because the ground will be able to absorb more heat than it would have otherwise.

Monday

Date
and Time .... Location ....

Relative Humidity .... Wind Speed and Direction ....
Atmospheric Pressure .... Cloud Cover..... oktas/overcast Temperature ....

Cloud Types ....

General Observations ....

Predictions for Tomorrow ....

Tuesday

Date
and Time .... Location ....

Relative Humidity .... Wind Speed and Direction ....
Atmospheric Pressure .... Cloud Cover..... oktas/overcast Temperature ....

Cloud Types ....

General Observations ....

Predictions for Tomorrow ....

Wednesday

Date
and Time .... Location ....

Relative Humidity .... Wind Speed and Direction ....
Atmospheric Pressure .... Cloud Cover..... oktas/overcast Temperature ....

Cloud Types ....

General Observations ....

Predictions for Tomorrow ....

Thursday

Date
and Time .... Location ....

Relative Humidity .... Wind Speed and Direction ....
Atmospheric Pressure .... Cloud Cover..... oktas/overcast Temperature ....

Cloud Types ....

General Observations ....

Predictions for Tomorrow ....

Friday

Date and Time .... Location ....

Relative Humidity ....   Wind Speed and Direction ....

Atmospheric Pressure ....   Cloud Cover..... oktas/overcast   Temperature ....

Cloud Types   ....

General Observations   ....

Predictions for Tomorrow   ....

Saturday

Date and Time .... Location ....

Relative Humidity ....   Wind Speed and Direction ....

Atmospheric Pressure ....   Cloud Cover..... oktas/overcast   Temperature ....

Cloud Types   ....

General Observations   ....

Predictions for Tomorrow   ....

Sunday

Date and Time .... Location ....

Relative Humidity ....   Wind Speed and Direction ....

Atmospheric Pressure ....   Cloud Cover..... oktas/overcast   Temperature ....

Cloud Types   ....

General Observations   ....

Predictions for Tomorrow   ....

Weather Predictor's Tip

The heat index, also known as the apparent temperature, is used by the National Weather Service in the United States to measure how hot it actually feels when relative humidity is added to the actual air temperature. The heat index is determined for temperature readings in the shade at wind speeds of 6 mph (10 km/h); exposure to the Sun can add up to 15°F (8°C) to the heat index value.

Monday

Date and Time .... Location ....

Relative Humidity .... Wind Speed and Direction ....

Atmospheric Pressure .... Cloud Cover..... oktas/overcast Temperature ....

Cloud Types ....

General Observations ....

Predictions for Tomorrow ....

Tuesday

Date and Time .... Location ....

Relative Humidity .... Wind Speed and Direction ....

Atmospheric Pressure .... Cloud Cover..... oktas/overcast Temperature ....

Cloud Types ....

General Observations ....

Predictions for Tomorrow ....

Wednesday

Date and Time .... Location ....

Relative Humidity .... Wind Speed and Direction ....

Atmospheric Pressure .... Cloud Cover..... oktas/overcast Temperature ....

Cloud Types ....

General Observations ....

Predictions for Tomorrow ....

Thursday

Date and Time .... Location ....

Relative Humidity .... Wind Speed and Direction ....

Atmospheric Pressure .... Cloud Cover..... oktas/overcast Temperature ....

Cloud Types ....

General Observations ....

Predictions for Tomorrow ....

1029   1027

880

890

1029

Friday

Date
and Time .... Location ....

Relative Humidity .... Wind Speed and Direction ....
Atmospheric Pressure .... Cloud Cover..... oktas/overcast Temperature ....

Cloud Types ....

General Observations ....

Predictions for Tomorrow ....

Saturday

Date
and Time .... Location ....

Relative Humidity .... Wind Speed and Direction ....
Atmospheric Pressure .... Cloud Cover..... oktas/overcast Temperature ....

Cloud Types ....

General Observations ....

Predictions for Tomorrow ....

Sunday

Date
and Time .... Location ....

Relative Humidity .... Wind Speed and Direction ....
Atmospheric Pressure .... Cloud Cover..... oktas/overcast Temperature ....

Cloud Types ....

General Observations ....

Predictions for Tomorrow ....

Weather Predictor's Tip

It can never be too cold to snow, but in very cold, dry conditions fewer clouds will form. Typically, very cold air contains little moisture unless the cold air blows from across a lake or ocean, where it can collect moisture. Most heavy snowfalls occur in temperatures 15°F (-9°C) or above. When there is a low dew point (the temperature at which the air saturates), the chance of snow increases.

Monday

Date and Time .... Location ....

Relative Humidity .... Wind Speed and Direction ....

Atmospheric Pressure .... Cloud Cover..... oktas/overcast Temperature ....

Cloud Types ....

General Observations ....

Predictions for Tomorrow ....

Tuesday

Date and Time .... Location ....

Relative Humidity .... Wind Speed and Direction ....

Atmospheric Pressure .... Cloud Cover..... oktas/overcast Temperature ....

Cloud Types ....

General Observations ....

Predictions for Tomorrow ....

Wednesday

Date and Time .... Location ....

Relative Humidity .... Wind Speed and Direction ....

Atmospheric Pressure .... Cloud Cover..... oktas/overcast Temperature ....

Cloud Types ....

General Observations ....

Predictions for Tomorrow ....

Thursday

Date and Time .... Location ....

Relative Humidity .... Wind Speed and Direction ....

Atmospheric Pressure .... Cloud Cover..... oktas/overcast Temperature ....

Cloud Types ....

General Observations ....

Predictions for Tomorrow ....

Friday

Date
and Time .... Location ....

Relative Humidity ....   Wind Speed and Direction ....
Atmospheric Pressure ....   Cloud Cover..... oktas/overcast   Temperature ....

Cloud Types ....

General Observations ....

Predictions for Tomorrow ....

Saturday

Date
and Time .... Location ....

Relative Humidity ....   Wind Speed and Direction ....
Atmospheric Pressure ....   Cloud Cover..... oktas/overcast   Temperature ....

Cloud Types ....

General Observations ....

Predictions for Tomorrow ....

Sunday

Date
and Time .... Location ....

Relative Humidity ....   Wind Speed and Direction ....
Atmospheric Pressure ....   Cloud Cover..... oktas/overcast   Temperature ....

Cloud Types ....

General Observations ....

Predictions for Tomorrow ....

Weather Predictor's Tip

A "super-typhoon" is a typhoon that reaches maximum sustained surface winds for one minute of at least 150 mph (240 km/h) based on a scale used by the U.S. Joint Typhoon Warning Center. This is the equivalent to a category 5 hurricane on the Saffer-Simpson scale or a category 5 severe tropical cyclone in Australia.

Glossary

absolute humidity The amount of water vapor in a given volume of air.

absolute zero The lowest possible temperature where no molecular motion takes place. It is calculated at -460°F, -273°C, or 0°K. Absolute zero has never been attained.

advection Transport of an atmospheric property, such as moisture, by the wind.

advection fog Fog produced when warm, moist air moves over a cold surface and the air cools to below its dew point.

air density Mass per unit volume of air; the density will change depending on temperature, altitude, and pressure.

air mass A large body of air with similar horizontal temperature and moisture that can extend hundreds or thousands of miles. When air masses meet, they can form fronts, which are narrow zones of transition.

air pressure (also atmospheric pressure or barometric pressure) The force exerted on a surface by the weight of the air above it. Pressure is calculated in a unit called a kilopascal.

altocumulus A mid-level cloud that typically occurs in layers or patches with wavy, rounded masses or rolls.

altocumulus castellanus An altocumulus cloud showing vertical development that often assumes the shape of a castle and is associated with precipitation.

altostratus A mid-level cloud composed of gray or bluish sheets of uniform appearance.

ambient air Air surrounding a cloud or a rising or sinking packet of air.

ambient temperature Temperature of the surrounding air.

anemometer An instrument to measure wind speed.

anticyclone A high-pressure area characterized by winds rotating clockwise and outward in the northern hemisphere and counterclockwise in the southern hemisphere. Anticyclonic rotation is associated with dry weather.

atmosphere The layer or layers of gas above a planet. On Earth the atmosphere consists of nitrogen, oxygen, and smaller amounts of other gases and is composed of five layers. The lowest layer is called the troposphere.

barometer An instrument used to measure air pressure in terms of units of measurement called the kilopascal. Barometers also measure pressure in millibars and inches of mercury. An aneroid barometer is one that contains no liquid.

Beaufort scale A scale used to measure wind strength. It was invented by Britain's Admiral Sir Francis Beaufort in 1805.

biosphere The region in which life occurs that includes the air, land, and water.

blizzard A severe snow storm lasting four or more hours characterized by low temperatures, strong winds, and poor visibility due to blowing snow.

centrifugal force A force directed outward, away from the center of a rotating object. It is equal in magnitude to the centripetal force but in the opposite direction.

centripetal force An inward-directed force that limits an object to a circular path. It is equal in magnitude to the centrifugal force but in the opposite direction.

cirrocumulus A high-level cloud that appears as a white patch of cloud.

cirrostratus A high-level cloud that often covers the entire sky.

cirrus A high-level cloud composed of ice crystals

that can appear in patches, filaments, or narrow bands.

chinook Relatively warm, dry gusty winds that can occur to the leeward side of mountain ranges. In parts of Europe these are known as Föhn winds.

climate The prevailing or characteristic meteorological conditions of a particular locale or region based on average weather conditions.

cloud A condensed form of atmospheric moisture containing tiny water droplets and/or ice particles.

cloud base The lowest portion of a cloud.

coalescence The collision or merging of water droplets into a single larger droplet.

cold front The leading edge of a cooler air mass.

condensation The process by which vapor becomes liquid or solid; the opposite of evaporation. Condensation is the process by which clouds form.

continental air mass An air mass that forms over land; it is normally relatively dry.

convective precipitation Precipitation originating from convective clouds, usually cumulonimbus or cumulus congestus.

convergence A large horizontal inflow of air into a region on the Earth's surface.

coriolis effect The force on air due to the rotation of the Earth, which causes it to turn to the right of its path in the northern hemisphere and to the left in the southern hemisphere.

cumulonimbus An exceptionally dense, vertically developed convective cloud, often with an anvil-shaped top. It is frequently accompanied by heavy showers, lightning, thunder, and sometimes hail.

cumulus A dense, well-defined cloud in the form of individual, detached domes or towers

with rounded tops and a flat base. Cumulus humilis clouds are often called fair weather cumulus, while cumulus clouds with vertical development are called cumulus congestus or cumulus castellanus.

cyclone A movement of air around a low-pressure area. The circulation of air is counterclockwise in the northern hemisphere and clockwise in the southern hemisphere. Cyclonic circulation can generate powerful storm systems, including tornadoes, hurricanes, and typhoons.

depression (also frontal depression) A system in which air masses of different temperatures and humidity meet and rotate slowly around a center of low pressure. It brings unstable, changeable weather.

dew Condensed water that forms on objects near the ground when their temperatures have fallen below the dew point of the surface air. Clear nights with cooling temperatures are most favorable for dew formation.

dew point The temperature at which air cools sufficiently to become saturated with moisture.

Doppler radar A type of radar that meteorologists rely on for identifying wind and moisture movements within clouds.

downdraft A small-scale column of air that rapidly sinks toward the ground, usually accompanied by precipitation, especially during thunderstorms.

downstream The direction in which wind is moving.

drizzle Precipitation from stratus clouds consisting of tiny, fine water droplets that are much smaller than in rain.

drought An extended period of dry weather.

El Niño El Niño is Spanish for "The Christ Child." A periodic warming of coastal waters off eastern South America around

Christmastime because of the trade winds. The phenomenon can last for several months and produces dramatic weather changes throughout the world.

evaporation The process by which a liquid changes into a gas. It is a crucial process in the water cycle as water from oceans, lakes, streams, and the ground is transformed into water vapor that rises into the atmosphere.

flurry Precipitation in the form of snow from a convective cumulus-type cloud that tends to be sporadic and light.

fog A cloud that forms near the Earth's surface. It usually consists of tiny water droplets and reduces visibility.

forecast A description of significant weather activity anticipated in the near future (usually no more than five days ahead) based on analysis of observations from many sources, including weather stations, satellites, radar, and weather balloons.

freezing rain Rain that freezes on impact to form a coating of ice on the ground.

front The leading edge of an advancing air mass that forms a boundary or transition zone between two air masses of different density and temperature. Fronts are named based on the air mass behind them, e.g., a cold front is named for a cold air mass.

frost Water vapor that forms a layer of ice on a surface colder than the surrounding air with a temperature below freezing.

Fujita scale (or F Scale) A scale to measure the strength of a storm, usually a tornado, based on the damage caused.

funnel cloud Cloud extending from the base of a towering cumulus characterized by a rotating column of air that, unlike a tornado, is *not* in contact with the ground.

gale A strong wind. A gale warning is issued for winds of 39–54 mph (63–87 km/h).

geosphere The solid earth including the continental and oceanic crust.

graupel Snow pellets.

greenhouse effect The warming and insulating effect due to the capacity of certain gases in the atmosphere to trap heat emitted from the Earth. This is a natural process without which the Earth would be too cold for human life. Scientists are concerned that the buildup of gases from burning fossil fuels may amplify the natural process, trapping too much heat.

gust A sudden, brief increase in wind speed that typically lasts less than 20 seconds.

hail Precipitation in the form of chunks of ice associated with thunderstorms. Hail ranges from pea-size to, occasionally, orange-size.

heat index An index combining air temperature and relative humidity to determine an apparent temperature to provide a better idea of how hot it actually feels.

high pressure An area of high atmospheric pressure usually associated with dry air. It is also known as an anticyclone.

humidity The measure of moisture or water vapor in the air. (See also absolute humidity and relative humidity.)

hurricane A tropical storm with wind speeds of 73 mph (117 km/h) up to 259 mph (414 km/h) that can extend for thousands of square miles. In the eastern Pacific, these cyclones are known as typhoons.

hydrosphere All bodies of water on Earth.

hygrometer An instrument used to measure humidity.

ice fog A type of fog composed of tiny suspended ice particles that forms at very low temperatures.

ice pellets Frozen water droplets in clouds.

instability Atmospheric conditions that favor the formation of thunderstorms. The instability causes rising packets of warm air to lift into higher reaches of the atmosphere rather than lose buoyancy, cool, and sink.

isobar A line on a weather map that connects points of equal pressure.

jet stream Relatively strong wind currents concentrated in a narrow stream about 9 miles (15 km) above Earth that can vary in position and orientation from day to day. The four jet streams exert a powerful influence on weather.

kilopascal The internationally recognized unit used to measure atmospheric pressure.

La Niña La Niña means "The Little Girl." A periodic cooling of waters in the eastern Pacific that causes dramatic changes in weather conditions throughout the world. It is the opposite of El Niño.

lake-effect snows Snowstorms that form on the downwind side of a lake in winter as cold, dry air collects moisture and warmth from the unfrozen bodies of water.

land breeze At night, the air over a coastal region cools faster than that over the ocean and causes the air to flow from the land toward the sea as warmer air above the sea rises.

latent heat Heat that is either emitted or absorbed by a unit mass of a substance when it undergoes a change of state, such as occurs during evaporation, condensation, or sublimation.

leeward Situated away from the wind, also downwind.

lightning A visible electrical discharge produced by thunderstorms due to the opposite charges of a cloud and the ground or of two clouds.

low-pressure area An area of low atmospheric pressure usually associated with cloud formation and precipitation.

Lows are also known as cyclones.

maritime air mass An air mass that originates over an ocean and usually brings moist air.

maximum-minimum thermometer A thermometer that measures the lowest and highest temperatures since the previous reading.

mesoscale A scale used to measure mid-sized weather systems of about 50 miles (80 km) to several hundred miles horizontally. The mesoscale examines phenomena smaller than those on the synoptic scale. A mesoscale would be used to measure a squall line but not a single storm cloud.

mesosphere The atmospheric layer between the stratosphere and the thermosphere and located 30–50 miles (50–80 km) above the Earth's surface.

meteorologist A person who studies and practices weather observation and analysis. There are many different types of meteorologists—e.g., research meteorologists, radar meteorologists, and climatologists.

meteorology The study of atmospheric conditions and their effect on weather patterns.

millibar (mb) A unit to express atmospheric pressure. Sea level pressure is normally close to 1013 mb.

mist Tiny water droplets suspended in a cloud over the ground. The water droplets are finer than in fog and reduce visibility less.

monsoon Seasonal winds that originate in the Indian Ocean and bring torrential rains to the Indian subcontinent and Southeast Asia.

occluded front A front that occurs when a colder air mass overtakes a warmer air mass.

ozone A form of oxygen that filters out ultraviolet rays of the Sun that can cause skin cancer.

polar air mass A cold air mass that forms at a latitude close to the poles.

polar easterlies Winds that blow from the high-pressure areas of the poles toward the low-pressure areas of the polar fronts at around 60°N and S.

precipitation Water that originates in the atmosphere and falls to the ground. It can take the form of hail, rain, freezing rain, sleet, and snow.

pressure gradient force A process in which air moves toward regions of lower pressure.

prevailing wind The wind direction most frequently observed during a given period.

psychrometer An instrument consisting of dry and wet bulb thermometers used to measure the water vapor content of the air.

radiation fog A type of fog produced over land where the Earth has emitted radiation (heat), reducing the air temperature to or below its dew point. It is also known as ground fog and valley fog.

radiosonde A package of instruments carried aloft by weather balloons that measures various atmospheric conditions.

rain Precipitation in the form of liquid water drops that have diameters greater than that of drizzle.

rain gauge An instrument used to collect rain and measure rainfall amounts.

relative humidity The percentage of moisture present in the air relative to the amount that would be present if the air were saturated (at 100 percent humidity).

ridge An area of relatively high atmospheric pressure. It is the opposite of a trough.

sea breeze Sea breezes develop during the daytime in warm sunny weather when the air over a coastal region heats more rapidly than that over the sea. As the warmer air rises, relatively cool air from the sea flows onshore to replace it.

sea level pressure (SLP) Mean atmospheric pressure at sea level.

shear The variation in wind speed and/or direction over a short distance. Shear usually refers to vertical wind shear, i.e., the change in wind with height.

shower Rain or snow from a cumuliform cloud, usually of short duration but often heavy.

sleet According to the U.S. definition, sleet consists of frozen raindrops that bounce upon impact with the ground. In other parts of the world, sleet is a mixture of rain and snow and occurs in a transition region between the two.

snizzle A frozen drizzle made up of extremely small particles of snow and ice that fall lightly.

snow Solid precipitation in the form of minute ice flakes that occur below 32°F (0°C).

squall line A line or band of active thunderstorms.

stationary front A transition zone that separates two different-pressure areas that remains in place up to a week and is associated with cloudy days and rainy weather.

stratiform cloud A cloud with extensive horizontal but little vertical development. These clouds produce precipitation that is usually continuous and uniform in intensity.

stratocumulus Low-level clouds that are arranged in rows, bands, or waves.

stratus A low, gray cloud layer with a fairly uniform base that may appear in ragged patches. Fog is usually a form of stratus cloud on the ground.

sublimation The process by which ice changes directly into water vapor without melting. In meteorology, it can also refer to the transformation of water vapor into ice.

supercooled Water (or another substance) that can become very cold (in a cloud, for instance) without freezing.

synoptic scale A meteorological scale that refers to weather systems with horizontal dimensions of several hundred miles or more, such as high- and low-pressure areas. (See also mesoscale.)

temperature The degree of hotness or coldness measured against some definite scale by means of a thermometer.

thermosphere The atmospheric layer above the mesosphere that extends from 50 to 430 miles (80 to 700 km).

thunderstorm A local storm, produced by a cumulonimbus cloud and accompanied by thunder and lightning.

tornado A violent funnel-shaped wind vortex in the lower atmosphere with upward spiraling winds of high speeds spawned by severe thunderstorms.

towering cumulus A large cumulus cloud with great vertical development, usually with a cauliflower-like form.

trade winds Winds that are prevailing in much of the tropics and blow from the subtropical high-pressure areas to equatorial low-pressure areas.

tropical air mass A warm, moist mass of air that forms in the subtropics.

tropical depression A system of thunderstorms and clouds often associated with a cyclonic wind circulation.

tropopause The upper boundary of the troposphere, usually characterized by an abrupt change in temperature with altitude.

troposphere The lowest layer of the Earth's atmosphere. It is where most weather occurs.

trough An elongated area of relatively low atmospheric pressure.

typhoon A hurricane that forms in the western Pacific Ocean, Indian Ocean, or in the waters near Australia.

updraft A small-scale packet of rising air. If the air is sufficiently moist, then the moisture condenses to become a cumulus cloud. Updrafts are a factor in thunderstorm activity.

upper-level divergence A decrease in a mass of air in the upper atmosphere, a phenomenon associated with thunderstorm creation.

upwind Refers to the source of the flow of air.

virga Streaks or wisps of precipitation falling from a cloud but evaporating before they reach the ground.

warm front The leading edge of a warm air mass that is often accompanied by cloud formation and precipitation.

waterspout A small, relatively weak tornado that occurs over water.

weather Day-to-day atmospheric conditions described in relation to temperature, pressure, humidity, clouds, wind, precipitation, and the effects of these characteristics on humans, animal life, and the environment.

weather balloon Large balloon filled with helium or hydrogen that is sent aloft to measure atmospheric conditions using a pack of instruments called a radiosonde.

westerlies Prevailing winds that flow from the southwest in the northern hemisphere and northwest in the southern hemisphere.

wind The horizontal movement of air relative to the Earth's surface.

wind chill A measure of the combined cooling effect of wind and temperature.

windward The direction from which the wind is blowing; also known as upwind.

Further Resources

Web Sites

On the science of meteorology:

www.theweatherprediction.com/habyhints

www.meteor.wisc.edu/~hopkins/100hold/wx-edlnk.htm

ww2010.atmos.uiuc.edu/(Gh)/guides/mtr/home.rxml

www.stuffintheair.com

www.refdesk.com/weath1.html

storm.uml.edu/~ammetsoc/wx_link/climo.html

www.wxqa.com/archive/obsman.pdf

temp-scales.org

Student and teacher resources:

delta-education.com/downloads/samples_dsm/738-6022Res.pdf
Student references and resources.

www.uq.edu.au/_School_Science_Lessons/UNPh37.html
School science experiments.

www.miamisci.org/hurricane/weatherstation.html
Miami Museum of Science make a weather station

Weather-tracking resources:

cirrus.sprl.umich.edu/wxnet/wxcam.html
Weather cameras across the United States.

asp.usatoday.com/weather/weatherfront.aspx
USA Today weather.

home.accuweather.com/index.asp?partner=accuweather
Accuweather for the United States.

www.bbc.co.uk/weather/weatherwise/index.shtml
International weather from the BBC.

www.weather.com
The Weather Channel.

www.theweathernetwork.com
Canadian weather information.

www.rmets.org
The U.K.'s Royal Meteorological Society.

www.ametsoc.org/amsedu/dstreme/
The American Meteorological Society.

Sharing your observations with other weather trackers:

www.wunderground.com
Worldwide weather resources and information.

www.cwop.net
Citizen Weather Observer Program.

www.weatherbug.com
Access to weather information from around the world.

www.nws.noaa.gov/om/coop
The U.S. National Weather Service's Cooperative Program.

Government weather services:

www.australiasevereweather.com/techniques
Weather observation tips from the Australian Severe Weather Service.

www.bom.gov.au
Australian Government Bureau of Meteorology.

www.met-office.gov.uk
The U.K. Meteorological Office.

www.msc-smc.ec.gc.ca/contents_e.html
Meteorological Office of Canada.

www.nws.noaa.gov
The U.S. National Weather Service.

www.nhc.noaa.gov
The U.S. National Hurricane Center.

Books

Aguardo, Edward, and Burt, James, *Understanding Weather and Climate*, Prentice Hall, 2006.

Buckley, Bruce, Hopkins, Edward, and Whitaker, Richard, *Weather: A Visual Guide* (Visual Guides), Firefly Books, 2004.

Burt, Christopher, and Stroud, Mark, *Extreme Weather: A Guide and Record Book*, W.W. Norton & Company, 2004.

Cantrell, Mark, *The Everything Weather Book: From Daily Forecasts to Blizzards, Hurricanes, and Tornadoes*, Adams Media Corporation, 2002.

Cox, John D., *Weather for Dummies*, For Dummies, 2000.

Dunlop, Storm, *The Weather Identification Handbook: The Ultimate Guide for Weather Watchers*, The Lyons Press, 2003.

Geer, I.W. (ed.), *Glossary of Weather and Climate*, American Meteorological Society, 1991.

Gibbons, Gail, *Weather Forecasting*, Aladdin, 1993.

Goldstein, Mel, *The Complete Idiot's Guide to Weather*, Alpha, 2002.

Flannery, Tim, *The Weather Makers: How Man Is Changing the Climate and What It Means for Life on Earth*, Atlantic Monthly Press, 2006.

Hodgson, Michael, *Basic Essentials Weather Forecasting*, Globe Pequot, 1999.

LeMone, Margaret, *The Stories Clouds Tell*, AMS, 1993.

Monmonier, Mark, *Air Apparent: How Meteorologists Leaarned to Map, Predict, and Dramatize Weather*, University of Chicago Press, 1999.

Nese, Jon, and Grenci, Lee, *A World of Weather: Fundamentals of Meteorology*, Kendall Hunt Publishing Co., 2005.

Rubin, Louis, Duncan, Jim, and Herbert, Hiram, *The Weather Wizard's Cloud Book: A Unique Way to Predict the Weather Accurately and Easily by Reading the Clouds*, Algonquin Books, 1989.

Sheets, B., and Williams, J., *Hurricane Watch: Forecasting the Deadliest Storms on Earth*, Vintage Books, 2001.

Sorbjan, Z., *Hands-On Meteorology*, AMS, 1997.

Streluk, Angella, and Rodgers, Alan, *Measuring the Weather: Forecasting the Weather*, Heinemann Educational Books, 2002.

Williams, Jack, *The Weather Book: An Easy-to-Understand Guide to the USA's Weather*, Vintage, 1997.

Index

advection 100, 155
advection fog 83
aerovanes 64–65
air
    density 167
    heat bursts 173
    water vapor 7
    wind 7
air masses 16–17
    fronts 22–25
    stable air 98–99
    unstable air 98
    weather prediction
        96–97
air pressure 14–15
    wind and 18–19
Alberti, Leone Battista 64
altocumulus 23, 74
    associated weather 81
altostratus 23, 74–75
    associated weather 81
anemometers 64
    home weather stations
        57
aneroid barometers 63
animals, folklore
        forecasting 110–111
anthropogenic climate
        26–27
anticyclones 15
apparent temperature 213
Aristotle 43
atmosphere 8–9
    circulation 18–19
    greenhouse effect 29
    ozone hole 29
atmospheric pressure 7,
        14–15
    weather prediction
        96–97
    wind and 18–19
Automated Surface
        Observation Systems
        (ASOS) 47
Automated Weather Sensor
        System (AWSS) 47
automatic weather stations
        (AWS) 47
avalanches 93

ball lightning 40–41
balloons, weather 48
barometers 62–63, 195
    aneroid 63
    digital 63
    home weather stations
        57
    invention 42
    mercury 63
    reading 62

barometric gradient 18
bats, folklore forecasting
        111
Beaufort scale 101
birds, folklore forecasting
        111
black ice 95
blizzards 93
blowing snow 93
Brinkhuis, Henk 45

"calm before a storm" 169
cats, folklore forecasting
        110
Celsius, Anders 58
Celsius scales 58
charts see weather maps
Chicago school of dynamic
        meteorologists 20
Chinooks 103
circulation, atmospheric
        18–19
cirrocumulus 72
    associated weather 81
cirrostratus 72–73
    associated weather 81
cirrus 22, 23, 72
    associated weather 81
cirrus aviaticus 73
    associated weather 81
Citizen Weather Observer
        Program 68
climate 26–27
    anthropogenic 26–27
    zones 27
climate change 30–31
    global change 28–29
    historical 45
clouds 7
    classifying 70–71, 75
    color 157
    convective 78–79
    coverage 153
    high-level 72–73
    low-level 76–77
    mid-level 74–75
    motion 187
    observing 115, 121, 137
    opacity 135
    predicting weather
        from 70–71
    stable and unstable
        conditions 99
    warm fronts 23
    water cycle 12
    weather and type of 81
    see also individual cloud
        types, e.g. cirrus
coalescence 86
coastal defenses 36

cold fronts 22–23
cold occlusions 25
computers
    forecasting 44–45
    supercomputers 50–51
    thunderstorms 90
condensation, water
        cycle 12
continental arctic air
        masses (cA) 16
continental polar air
        masses (cP) 16
continental tropical air
        masses (cT) 17
contrails 73
    associated weather 81
convective clouds 78–79
convective precipitation 87
cooling system, clouds
        and 71
Cooperative Observer
        Program 68
Coriolis force 18
cows, folklore forecasting
        110
Cray, Seymour 51
crickets, folklore
        forecasting 110
cumulonimbus 79
    associated weather 81
    thunderstorms 89, 90
cumulus, waterspouts 41
cumulus castellanus 79
    associated weather 81
cumulus congestus 78
    associated weather 81
cumulus humilis 78
    associated weather 81
cumulus mediocris 78
    associated weather 81
cycles, 11-year 21
cyclones 15
    tropical 34–35

Daniell, John Frederic
        60
data
    sharing 68–69
    weather stations
        46–47
day length, equinox 11
De Cusa, Nicholas 42, 43
depressions, frontal 19
dew point 95
digital barometers 63
digital thermometers 59
doldrums 19
Doppler, Christian 49
droughts 30, 37
    driest location 31

Earth
    tilt 10–11
    water coverage 12
El Niño 31, 38–39
El Niño-Southern Ocean
        Oscillation (ENSO) 38
electrical fields, St. Elmo's
        fire 40
electronic weather
        stations 69
ENSO 38
environmental satellites 48
equinox 11
equipment
    home weather stations
        56, 57
    weather satellites 48
    weather stations 46
evaporation 12
exosphere 9
extreme weather 28–29
    records 31

Fahrenheit, Daniel 43
Fahrenheit scales 58
flash floods 36
flies, folklore forecasting
        110
floods 36–37
flurry, snow 93
fog 82
    freezing 95
    ice 95
    types of 83
    warm fronts 23
Föhn 103
folklore, forecasting
        108–109
Folli, Francesco 42
forecasting
    checklist 105
    folklore 108–111
    history 42–43
    modern 44–45
    probability 85, 163
    supercomputers 50–51
    tips for 104–105
fractocumulus 77
    associated weather 81
freak weather 40–41
freezing fog 95
freezing rain 87
frontal depressions 19
fronts 22–25
    cold 22–23
    gust 25
    occluded 24–25
    stationary 24
    warm 23
    weather maps 53

weather prediction
96–97
frost, predicting 95
frost lines 125
Fujita scale 32
funnel clouds 41, 82, 201
associated weather
81

Galileo Galilei 43
geostationary satellites 49
global climate change 28
Global Earth Observation
System of Systems
(GEOSS) 50
global warming 29
floods 36–37
Goethe, Johann Wolfgang
von 62
greenhouse effect 29
groundhogs, folklore
forecasting 110–111
Gulf Stream, global
warming 29
gust fronts 25

hail 87, 94
heaviest hailstones 31
heat 7
heat bursts 173
heat index 213
heat lightning 175
high pressure 15
temperature 203
weather prediction
96–97
wind 18
higher atmosphere 8
horse latitudes 19
Howard, Luke 75
humidity 7
maps 106
see also hygrometers
hurricanes 15, 34–35,
189, 193
Hurricane Katrina 30
increase in 30
naming 35
hygrometers 60–61
constructing 61
home weather stations
57
invention 42

ice
falling 40
predicting 94–95
ice crystals, rain 7
ice fog 95
ice pellets 94

ice sheets, global warming
29
ice storms 87, 147
instruments
home weather stations
56, 57
weather satellites 48
weather stations 46
Internet 68–69
using 107
isobars 53

Jeanne, Tropical Storm 31
jet streaks 21
jet streams 20–21

Katrina, Hurricane 30
Khansin 103
Kyoto Treaty 29

La Niña 38–39
lake effects 149
land breeze 102
landspouts 197
Lavoisier, Antoine Laurent
43
Leonardo da Vinci 60
lightning 89, 165
ball lightning 40–41
heat lightning 175
St. Elmo's fire 40
logs 59
low pressure 15
temperature 203
tropical cyclones 34–35
weather prediction
96–97
wind 18
lower atmosphere 8

mammatus 80
associated weather 81
maps see weather maps
maritime polar air masses
(mP) 16
maritime tropical air
masses (mT) 17
mathematical models
forecasting 44–45
supercomputers 50–51
thunderstorms 90
maximum-minimum
thermometers 59
Mean Sea Level Pressure
(MSLP) 14
measurements, air
pressure 15
mercury barometers 63
mesocale 104
mesohighs 97

mesolows 97
mesoscale 97
mesosphere 8
meteorology
history 42–43
modern 44–45
mist 82
Mistral 102
models
forecasting 44–45
supercomputers 50–51
thunderstorms 90
monsoons 37
moon, red 141
Morse, Samuel 43
mountain winds 102

nature, folklore forecasting
108–109
night length, equinox 11
nimbostratus 76
associated weather 81
Numerical Weather
Prediction (NWP)
44–45

oceans 12
climate change 30
El Niño 38–39
La Niña 39
upwelling 38
water circulation 19
oktas 71
Ooishi, Wasaburo 20
ozone hole 29

Palmén, Erik 20
pimpernels, folklore
forecasting 108
planetary winds 20
polar easterlies 15, 19
polar fronts 19
polar orbiting satellites 49
popping ears 14
precipitation 7
climate change 30
climate zones 27
cold fronts 22–23
convective 87
intensity 86
maps 106
predicting 84–85
process 86–87
rain gauges 66–67
types 87
water cycle 12
see also rain; sleet; snow
precipitation fog 83
prediction
checklist 105

folklore 108–111
history 42–43
modern 44–45
probability 163, 85
supercomputers 50–51
tips for 104–105
pressure 7, 14–15
making readings 62
weather maps 53
weather prediction
96–97
wind and 18–19
see also barometers
pressure anemometers 64
pressure gradient 18
prevailing westerlies 19
probability forecasting
85, 163
psychrometer 60

radar 49
radiation fog 83
rain
freezing 87
meaning 199
most in one day 31
predicting 84–85, 117,
119, 127
process 86–87
raindrops 145
rainfall statistics 67
water cycle 7, 12–13
rain bands 163
"rain before wind" 109
rain cascade 117, 177
rain gauges 66–67
home weather stations
57
"rain long foretold" 109
raindrops, shape 161
rainfall
climate change 30
drought 37
El Niño 38
monsoons 37
readings
locations 47
weather stations 47
"red sky at night" 108
relative humidity 60, 207
Richardson, Lewis 44
ridges 96
roll clouds 80
associated weather 81

Sahara dust 40
St. Elmo's fire 40
Santa Ana winds 103
satellites 48–49
satellite imagery 107

scales, meteorological 104
screens 56–57
sea breeze 102
sea gulls, folklore
    forecasting 110
sea levels, global warming
    29
seasons 10–11
shelf cloud 82
    associated weather 81
shelter, weather stations
    56–7
showers 199
Sirocco 103
sleet 87
    predicting 84, 94–95
sleet storms 94
smoke rising 109
snizzle 94
snow 215
    facts 93
    greatest fall 31
    ground temperature 211
    measuring 46
    measuring rain
        equivalent 143
    predicting 84, 92–93
    types of snowfall 93
snow boards 57, 67
snow flurries 93
snow grains 93
snow pellets 93
snow squall 93
snowflakes 87
    largest 93
software 69
solar activity, jet streams
    and 21
sonic anemometers 64
squall lines 90
squirrels, folklore
    forecasting 111
stable air 98–99
stationary fronts 24
steam fog 83
Stevenson screens 56–57
storms
    climate change 30
    gust fronts 25
    hurricanes 34–35
    occluded fronts 24–25
    see also thunderstorms
stratocumulus 76–77
    associated weather 81
stratosphere 8

stratus 76
    associated weather 81
stratus cloud 23
sublimation, frost 95
summer, increasing
    sunlight 11
Sun
    and Earth's tilt 10–11
    heat from 7
    jet streams and 21
    wind and 19
sunrise 10
    equinox 11
super-typhoons 217
supercells, tornadoes
    32–33
supercomputers 50–51
symbols, weather maps 53
synoptic scale 104

temperate zones 27
temperature 7
    climate change 30
    climate zones 27
    conversion 58
    greatest range 31
    greenhouse effect 29
    highest record 31
    hottest location 31
    lowest recorded 31
    maps 106
    mesosphere 8
    stratosphere 8
    thermosphere 9
    see also thermometers
thermometers 58–9
    constructing 58–9
    invention 43
    location 57
    scales 58
    types 59
thermosphere 8–9
thunder 89, 165
thunderstorms
    anatomy of 88–89
    calm before 169
    classification 129
    distance 91
    gust fronts 25
    predicting 90–91,
        133
    safety 171
    tornadoes 32–33
tilt of the Earth 10–11
Tornado Alley 32, 209

tornadoes 32–33, 185
    waterspouts 41
Torricelli, Evangelista 42
trade winds 19
tropical cyclones 34–35
tropical depressions 15
Tropical Storm Jeanne 31
troposphere 8
    global warming 29
troughs 91, 96, 151
twisters 32–33
typhoons 15, 34–35
    increase in 30
    super 217

unstable air 98–99
upper air, maps 107
upper-level divergence 96
upslope fog 83
upwelling 38

valley fog 83
valley winds 102
Vendavales 102
vertical scale 104
virga 159

wall cloud 82
    associated weather 81
warm fronts 23
warm occlusions 25
warming system, clouds
    and 71
water
    droughts 37
    evaporation 12
    floods 36–37
    molecules 139
    water cycle 7, 12–13
        building a model 13
        model of 13
water vapor 12, 42
waterspouts 41
weather
    climate and 26–27
    extreme 28–29, 31
    forecasting 42–43
    freak 40–41
weather balloons 48
weather charts, wind
    101
weather logs 59
weather machine 6–7
    atmospheric pressure 7
    temperature 7

water cycle 7
wind 7
weather maps 46
    humidity 106
    precipitation 106
    reading 52–53
    symbols 53
    temperature 106
    thunderstorms 91
    upper air 107
    using 106–107
    weather 46
weather radar 49
weather satellites 48–49
weather stations 181
    amateur 68–69
    automated 47
    electronic 69
    location 47, 57
    professional 46–47
    setting up 56–57
    taking readings 47
Weather Underground 69
weather watches 191
WeatherBug 69
wedge gauges 67
whirlwinds 41
wind 7, 18–19
    Beaufort scale 101
    causes 18
    direction 18, 100–101
    fasted recorded 31
    global pattern 19
    gust fronts 25
    jet streaks 21
    jet streams 20–21
    local 102–103
    prediction 179
    speed 100–101, 123,
        131, 205
    thunderstorms 89
    tornadoes 32–33
    tropical cyclones 34–35
    weather maps 53
    weather prediction
        96–97, 100–101
    whirlwinds 41
"wind from the north" 109
wind vanes 64–65
    constructing 65
winter, decreasing sunlight
    11
wooly bear caterpillars,
    folklore forecasting
    111